James John Garth Wilkinson

The Soul is Form and Doth the Body Make

The Heart and the Lungs, the Will and the Understand

James John Garth Wilkinson

The Soul is Form and Doth the Body Make
The Heart and the Lungs, the Will and the Understand

ISBN/EAN: 9783744755207

Printed in Europe, USA, Canada, Australia, Japan

Cover: Foto ©berggeist007 / pixelio.de

More available books at **www.hansebooks.com**

The Soul is Form and doth the Body make

The Heart and the Lungs
The Will and the Understanding

Chapters in Psychology

BY

JAMES JOHN GARTH WILKINSON
FELLOW OF THE ROYAL GEOGRAPHICAL SOCIETY

JAMES SPEIRS
36 BLOOMSBURY STREET, LONDON
1890

TO

OUR CHILDREN AND GRANDCHILDREN

THIS SMALL WORK IS

LOVINGLY DEDICATED

PREFACE.

BEGINNING at the end, I would prepare the Reader to expect, that if the heart, as a part of my thesis is, corresponds to the love or will of the person, it must be the most complex of fleshly organs. And not only the most complex, but on both sides the most communal, every part of it in the possible currency of every other part, as all man's feelings intercommunicate, and can feel one another. The heart of no animal can be set in parallel with the human heart. Pleasure of existence, appetite for maintenance, propagativeness, love of offspring, love of congenial places, organic actions and passions, are some general statement of animal natures and motives. And these are instinctive. But another kind of natures or "loves" belongs to a man. Granted that by correspondence many analogies in heart-character are suggested in the animal world as a whole set over against human nature; yet each single man overlaps in central qualities all animal natures put together. This is because of his complex constitution, its height and depth. And then because he has a free will and a free understanding, a free determination of his own; whereby he can use his "capacious powers" as he chooses; govern them; change their predominances; and subject them to incessant modifications. He is a field of his own experiments. Animals undergo experiments, but do not institute them. The fate of their instinct can be handled by man, the experimenter;

but it continues to be instinct however newly it is shaped.

You never speak of animal love or loves except in metaphor: it is common to speak of them as "nature." With no shock to exactness you speak of human love and loves both good and evil. Self-love is a fundamental heading. And when you consider human loves or determinations of liking, books are from them and about them, but cannot contain them. Genera, species, single specimens of affection, all these sizes englobed in human nature, what a world, what an abyss, what a record it is! Palæontology and Zoology shrink into nothing before it.

Now to come nearer to this heart, will, love, choice of good or *summum bonum*, though in the briefest abstract. First note ambition, self-management, in whatever degree, whether only to be and exist, or to rule. Then marriage-love, love of our children, love of home and its preservation, love of society and of our place in it; of maintaining that place or improving it. Friendship and sympathy with our friends. Love of things intellectual, of art, culture, science. By God's mercy possible love to God and love to man. And these loves, each of masterful capacity, subordinate to the best and highest of them, or not subordinate. And this manifold life,—"the love is the life,"—working ever onwards, and represented in the body of the heart, in its channels and currents. These latter at last become the works and ways of it. Now then think of the heart as a swarming city of mankind, and you will better realize what your heart carries. In that metropolis of interests your good ambition, indispensable for large uses, leads down fountains of healing waters into noble service: your base selfhood pillages the settled state for the palace of your lusts, and makes it the example of a general pollution and a profane dominion. The obedient flesh hardens or softens accord- to opposite loves; it becomes heart of flesh, or heart of stone. In less cardinal ways your resolves, good or evil, are represented in ramparts of flesh, and in blood that stands its ground and does defiance and faces reverse; your weak compliances in wavering blood and trembling caverns of heart. Wild generosity, too, lives here, and drains its sanguine purses; and avarice starves its own flesh, and kills its blood in grasping it. One love balances itself against another, and the remainder is then in force. But always you have commands firmamental above; and desperate fortunes below in their chambers of imagery. In short, the virtues and vices as tendencies, and then as wills, and then as additions to inherited nature, and then as bequests to children, are present and represented in your heart.

By some knowledge of the anatomical heart, and a belief in the divine Science of Correspondences, Psychology can now take you by the hand, and lead you in the bosom of the mind to a temple of heart, the building of the Divinity for Himself to dwell in; to catacombs of heart, heathen and Christian, the ghostly resorts of a line of ancestral natures ; and into the present arena where the wrestle and race of your day goes on. Knowledges for use lie in following the daylight first, and then the torchlight of Psychology, into these arcane places, most mysterious where the light is strongest. The Psychology or Soul-science with the Anatomy or Form-science, is in one sense, *Nosce teipsum*, know thyself. Through commissioned Swedenborg a scientific equivalent is added to self-knowledge, and holds up our nature's central mirror for a new *Confessio Amantis* or Lover's Confessional. In some future age it will be known what a help to spiritual insight is given in the position, that "the heart corresponds to the love and the lungs to the understanding."

February 18, 1890.

CONTENTS.

	PAGE
I. PHYSIOLOGY AND PSYCHOLOGY,	1

II. THE LUNGS—

General Remarks on the Pulmonary Circulation,	24
The Breathing Lungs and the Beating Heart,	28
Attractive Freedom a Foundation in the Body,	30
The Bronchial Vessels before and after Birth,	33
The Mission of the Lungs,	41
The Passage to Psychology,	43
The Will and the Understanding,	46
The Mental Side of the Correspondence,	68
The Separation of the Intellect from the Will,	74
Endowments from the Separation,	76
Further Bronchial and Pulmonary Considerations,	79
A Question from the Spiritual Side,	82
More about the New Will in the two Organisms,	84
Some Terms explained,	88
The Doctrine of the Human Body,	94
Old-World Psychology,	108
Weighing and Pondering,	113

III. THE HEART—

	PAGE
Centrals for the Centres,	125
The Coronary Vessels of the Heart,	130
Representation,	138
The Mental Heart,	147
The Circulation of the Heart and the Coronaries: their Correspondences,	163
Relations of the Brain before and after Birth,	168
Swedenborg on the Coronary Vessels of the Heart,	172
The Veins,	186
Psychology of the Veins,	200
The Capillaries and the Vasa Vasorum,	203
Nutrition,	209
How the Organs feed,	213
Arrested Physiology,	215

IV. PSYCHOLOGICAL NOTES—

Revealed Psychology,	222
A Word from Greek Philosophy—Clinamen,	238
Matter and Life,	246
A Door to Psychological Chemistry,	256
The Vital Principle,	262
The Elevation of Faculties,	268
Psychical Epochs,	271
The Word in its Epochs,	299

POSTSCRIPT, ... 315

THE SOUL IS FORM AND DOTH THE BODY MAKE.

I.

PHYSIOLOGY AND PSYCHOLOGY.

1. A CHILDHOOD with no expectations stands at the beginning of many departments of experience and knowledge. The man who first tasted animal food was a man of a tentative and curious genius, but little foresaw the vast creation of flocks and herds with which the appetites of his followers would cover the plains. He who first opened and peered into the dead body was of an audacious curiosity, and had no forecast of the sciences that would issue from his inquisitiveness. The

A

special origin of anatomy and physiology is unknown; and it is uncertain when these branches as sciences were introduced to medicine. The healing art no doubt long preceded them, and still subsists in rude races without any reference to diseased parts or functions. Herbalism, magic, astrology, and incantations are general Allheals which do not look towards science.

2. But as medicine grew, and emerged from what it now considers as the superstitious epoch, anatomy and physiology became more serious studies, the impetus of cure as a purpose was added to them, and the earnest dissection of the organism was inaugurated. We may probably credit physicians and surgeons with being the originators of anatomical science properly so called. They need not have been devoted practitioners of medicine, for the science they pursued was fascinating enough for its own sake to be a

life-occupation. Yet because of their alliance with medicine we may say that in early times, as now, anatomy and physiology were and are in their medical era.

3. Contemporaneously with this, they have become greater and more general sciences, prosecuted as knowledge on their own account. Their leading experts would not be content to tie them down to the use and service of the sick man. Medicine and Surgery want but a small part of their voluminous and ever-increasing informations. Regional divisions, big organs, leading functions, the skin and the senses, and the skeleton, are essential to be known for practice. These obvious parts of man, and common sense conversant with them, are the assistants of healing. The question might some day arise, if any one were so perverse as to put it, Whether medicine with its great modern learning,—the best and most success-

ful medicine,—makes actual use of much more than this modicum of anatomy and physiology? And furthermore, whether the conscious prepossession of these sciences unlimited, does not squander practice over an alien field, and divert from effect the simplest strokes of healing?

4. At all events we are now in a physiological epoch beyond the medical era, and the leaders have in view other objects than the treatment of disease, or even than the extension of surgery. The medical monopoly of physiology as connected with these arts is at an end.

5. From the first, or say from Aristotle, still himself a physician, anatomy and physiology have been "steeped" in one or all of the current philosophies. They are, as we have endeavoured to show, the continents and lower limits of all philosophy; the body, which is their subject, basing and

containing the mind, and, with this inhabitant, being the natural man.

6. Together with the medical era there is therefore the philosophical era of anatomy and physiology. This is diverse according to the rule of the philosophy. We may believe in the living body, in all that the senses reveal to the anatomist, as the whole account of the personality ; and fully admitting the mind in discourse, regard it also as a function of the body, and as having no existence separate from the matter of the brain. Broadly, then, you have materialistic physiology.

7. In the process of this, atheistic physiology, with a corner of pantheism for the prudent, arises with time and season. It is always waiting to come, and is a hungry form. Unlike practical medicine in having small requirements of knowledge of anatomy, atheistic physiology demands to register for

its account and cause the whole body of visible and ponderable man ; to know him from his seed to his stature. It dare not leave any of him out, lest other claims to possession should be pleaded. In at length finding out all, it can at last attain the crowning right of negation ; and declare authoritatively what is not, and veto presumed nonentities when they are incontinently pleaded. This is a proud throne, and boundless knowledge is the only legitimate kingdom and circumstance of it. We owe it something in science, for hard labour seven days a week is its royal doom. Its mind is "the supreme being" of the mental order.

8. Theism also is an old philosophy. But because its god, though helpful to many minds in the past and the present, is an unknown quantity and quality, it has no physiology special to it except from the suggestion, that the wonderful design of the

body must imply a wise designer and creator. Such true common perception, however helpful to the pious theist in common with the Christian, points not otherwise organically than that superhuman contrivance is demonstrated in the body. This fact, so far as it goes,—and the shortcomings and evils of human constitutions limit it for critical minds,—is a dictate of common sense, and not an instrument of physiological research, still less of psychological intuition. To mere science the god of the factory of the human machine is an arbitrary designer, and might have made it otherwise. The history of pious literature shows that the argument from design, Paley's argument, has not initiated science, has not added light to theology, and has not been impressive against atheism, or prosperous to religion. Granting design, which we must do, we may say that theistic physiology is, as a separate branch, a germ that can have no development, because the designer is a varying creature of theistic faith, unrevealed excepting to piety ; and such piety has no commerce with science. Whatever detailed physiology arose in early ages was independent of theism.

9. On the other hand, a real God, a Divine Man, has inevitably a divine physiology, and all organic forms and natures are derivations from it or shadows of it. A divine *neutrum*, a divine It,[1] is, according to the degree of culture, either a stock or a stone, or a formless will, or total impersonal nature, of which physiology knows nothing. From non-physiologicals physiology can never come, any more than love except from love,

---

[1] "The Hindu eagerly tells you that he believes in the Supreme—the one only God, as you do. But he speaks of the Supreme, not as He, but It; as an impersonal unconscious energy, a characterless self."—Dr. Pope in *The Missionary Field*, October 1889.

or wisdom except from wisdom. How It should ever be called He by our good theists is explained by the mercy of God, who breaks men's logic when for their sakes it is good that it be broken.

> " Deep in unfathomable mines
> Of never-failing skill,
> He treasures up His bright designs,
> And works His sovereign will."

10. We now come directly to Psychology. Some survival of its Most Ancient and Ancient perceptive and then intellectual intuitions and knowledges, is preserved in classic literature, as we see in John Gower's transcript of the Aristotelian teachings. But this goes no higher than to the inhabitation of man's passions in specific organs of the body. To Epicurus, and his disciple Lucretius, the brain is an uninhabited district. Neither do the passions inhabit structurally

or conveniently because there is no perception of the correspondence between the mind-man and the body-man. Psychology, therefore, as a Word about a soul, in the sense in which we now can use it, has no place as yet in this series of investigations concerning the body. The reason is, that no one knows anything about the soul as a reality commensurate with the body; and when this is the case, in point of science, the soul is a word, and the body a substantial reality. Now no undefinable word, no word labelling the unknown, can represent to the intellect anything that can influence a corporal being bodily, still less organically. A knowledge of the soul is necessary to enable us to perceive any relations that the body can have with it. Relations of analogy even require this, for similes cannot run between two things in which the one is all of outline and the other is shapeless. Christians do indeed believe in the soul, but the matter here concerns science, and co-real knowledge of soul and body is implied in knowledge of their intercourse.

11. The scientific world has not then at present attained to any psychological physiology.

12. We have thus proceeded, we cannot say, ascended, from the medical and utilitarian or professional physiology, to the philosophical physiology, which places knowledge, science, penetration into nature, thus mental culture and development, before healing. The advanced schools of vivisection and naturalism boast that knowing is their high aim and end, and that use, and service to man otherwise, is a secondary if not a spurious consideration. They are so far right, that their unlimited parts consist of " useless scientifics" which concern only their own academic interests.

13. What remains? By the consent of all schools, and especially of the materialist and atheist school, Freedom remains for the sake of *modus vivendi*. On its area, consecrated by martyrs of all kinds and creeds, by Servetus and Galileo and Giordano Bruno and many others, we may dare to say that Theological Physiology remains, and can stand up without shame or fear in our world of great permissions. Swedenborg was commissioned to bring it into the world. And through his agency Psychological Physiology is definitely founded on an even or equated knowledge of the body and the soul: the soul being a complete and in nowise a disembodied man within the known embodied man, and having therefore a superiorly substantial or spiritual physiology.

14. Such things are a Revelation: in this case to the rational mind of a man, and through him to the whole climate of human

rationality. One man with an age girt about him, and a spiritual world inspiring him, begins everything. So Plato began his work, and so Newton his. So also Swedenborg began his. He was then taken up for a higher knowledge and use to man, indispensable to the rounding and perpetuation of the lower knowledge and use, and he was instructed specifically on the superior plane by the Lord. Not Knowledge for its own sake, but Use, service to man, usefulness, was his only sublimity, his only poetry, the keynote of the august and immeasurable spirit of his science. Practical good shut up the abyss which had swallowed so many ages. His reason, his rational mind, thus made initiative as one mind at first, was still the limitaneous field of the divine instruction. Rational is its name for ever. "*Nunc licet intrare intellectualiter in mysteria fidei.*" Now may we enter intellectually into the mysteries of faith. Thus theology has become a reasonable field of knowledge, and Faith is perceived holy Truth. The boundary of the knowledge for each man is the regeneration of the daily life, for this is the condition of letting down and drawing up at the well of the living waters. Upon new ages of love and its life hang therefore the progress and continuance of the higher and highest sciences. They descend from heaven into minds heeding of heaven. Angels they are, but only ministers to practical good.

15. In the descending series, from a rational theology, from one God-Man in one Person, Christ the Lord, Psychology is born, and it will be the parent of a radiant physiology. What such knowledge can be is too wonderful for us. It will, however, as now, be human first, though attended by Creation. It will be organic, as every man and woman is an indivisible organ (Atomos). It will be

limited without by the tenderness and loyalty of reason, so as to be boundless within. And it will serve True Christian Religion as a divinely instructed handmaid.

16. This is written in Genesis. God made man in the Image and Likeness of God. He breathed into his nostrils the breath of lives, and man became a living soul. A living soul according to the image and likeness of the Maker! Here is theological psychology in its origin: God - Man and Adam - man. Not unknown either term, because Revelation makes both terms known. The archetype is given to holy knowledge in the image, and the image in the Archetype. Man is the finite human form because God is the infinite human form. His creation is the descent of the Infinite humanity into the finite organism; no arbitrary creation which could be otherwise, but the influx of the divine Man and Father into the accom-

modated correspondent human. The Father, God, in His form can have no other child than man in his form; Man, His image and likeness.

17. There is therefore a divine Physiology or Psychology,—call it by either name,—from the love of God to man, and from the likeness of man to God. By the latter condition, man, the image, if he will, can imagine and image God, and man, the likeness, can be like, that is, can love God.

18. This divine or theological psychology is essentially doctrinal, for unless there be true or rational doctrines concerning God, nothing of Him is known, and it is in the knowledge of God that He can be correlated to His image and likeness, Man, and in this be the revealed subject of a theological psychology.

19. In the same way there is an image-and-likeness psychology in man himself;

for the soul is the archetype of the body, as God is the archetype of the soul. It builds the body, image-wise and likeness-wise, and similars flow into similars when it lives and loves and works in the shapes which are its children and children's children. Correspondence is the name and sanction between all the lineages of the patriarchate in all its tribes, and in all their places in the temple. All are images and likenesses, and not otherwise legitimate creations.

20. Our little science of physiology, standing like the Israelites under Sinai at the base of the mountain of these correspondences, awaits its commanding Soul from the new psychology, and its life from the Divine Humanity in its now Rational Annunciation.

21. There is besides an incorporation and physiology of humanity attested since the earliest times. Heads have always been in name and fact. In all nations, kingdoms and

bodies of men, in all houses, there are both heads and hearts, which are in the offices of ruling conjointly by whatever wisdom and love is alive in their ages or generations. If there are heads and hearts, there are also humanitary senses; hands and feet; in short, all the organs on a social, national and integral scale that belong *in minimis* to the private man. There is the social body, the body politic, and the like. All societies for whatever object banded are bodies of men. Each society is one greater man. There is therefore a greatest man, more or less consistently organic according to his virtue, as the reason and rule of the creation and existence of every planet. The existence of reacting individuals, houses, societies, nations, can give rise to nothing but the organization of the individuals; and every individual is a human form. So all humanity is no promiscuous mankind, but a Man of men; a human form. This is a necessary fact of thought, and not an accidental metaphor in language. History is the unconscious exponent of it, and lives within it.

22. The home and pressure of this World-Man; this human form communicated and lent to our universal and general corporeal estate, without which pressure its warring sections would not cohere, is in the heavens and in the spiritual world. Heaven is MAXIMUS HOMO, the Greatest Man,—Man gathered up from all ages and worlds in his grandest Humanity. And so it is that all in heaven, and all who are preparing for their use in heaven, are in some province of the Greatest Man. They are functions or Uses, loving men or working men according to the organs in which they live and work. They are angels of the head, or of the heart, or the lungs, or the bones, or the hands, or the feet, or of some of the other parts, or organs.

They serve, by divine Providence, the men and women here who are in their own images and likenesses, and who die out of the deadness of the body into the life of the spirit; claiming them for the organs to which they will belong. For the whole natural world is the stomach which by the first digestion of death feeds the heavenly appetite of the immense other life with new associates. There is therefore a further physiology here: the physiology or truly the psychology of the spiritual body called humanity. It subsists from the fact that God from the origin is a Man, builds His heaven in His image and likeness, and operates all creation from His own Manhood. Swedenborg has been permitted to open this Pneumatology also,—this doctrine of spirits,—in his *Arcana Cœlestia*, and has treated of the characteristics of those who are in the various organic provinces of the priesthood and royalty above.

But also of those who exist as the disorders and malformations opposed to the divine form and uses of the organs.

23. But what is the first Soul, the Psyche? It is the inmost degree in every man precedent to his creation; for it is the Lord's abode with him; it is holy, unassailable, and eternal; a human divine projection. It may be called by eminence the soul of the soul. The finite possible man consists of three degrees below it. These follow in order, and a large common sense, or as the word may be, good sense, can comprehend their name and distribution. First, there is a celestial degree, involving love with its wisdom as supreme and all - determinant. Second, there is a spiritual degree, involving intellect and its conscience as absolute and commanding. Third, there is a natural degree containing both the former degrees, the obedience and *obsequium* of both; receiving their love and

intelligence not in their essences but in their external representations. Man is born into the worldly natural degree, a lower plane still, a spirit clothed upon with matter, space and time. According to his life here, and the character which that life makes him, he comes after death, by a shorter or longer way, into the degree which he has opened internally by his career in this world; by "the deeds done in the body." If he comes finally into the natural degree, the others are closed to him, and he knows them only representatively, but not as they are in themselves. If he comes into the spiritual degree, the celestial remains above and beyond him. In whatever degree he lives, it is filled to his utmost capacity of bearing with the happiness of heaven. But if his life is of confirmed evil or selfishness, only the natural sensuality of him is left, and he is inverted, with the lowest ends in rule, and the highest in subjection. A small epitome of our Psychology.

24. We complete the golden words of Edmund Spenser in his *Hymne in honour of Beautie* :—

> "So every Spirit as it is most pure,
> And hath in it the more of heavenly light,
> So it the fairer body doth procure
> To habit in, and it more fairly dight
> With cherefull grace, and amiable sight.
> For of the Soul the body form doth take;
> For soul is form and doth the body make."

II.

The Lungs.

General Remarks on the Pulmonary Circulation.

25. In translating Swedenborg's wonderful Work, *Angelic Wisdom concerning the Divine Love, and concerning the Divine Wisdom*, I longed to enjoy rational insight of his doctrine that the bronchial arteries and veins give the lungs the faculty of respiratory motion, and enable them herein to be separate from, or non-synchronous with, the pulsatory motions of the heart. If we accept the doctrine, a physiological *arcanum* lies here. Perhaps the above formula is not incompatible with the current scientific account. According to the latter, the pulmonary

arteries and veins when opened at birth pass the blood through the lungs from the right ventricle of the heart to its left auricle, in order to its purification and aëration in the capillaries of the vesicular lungs, and to its conversion there from venous into arterial blood, which is then served out to the whole system by the left ventricle through the aorta. The account of the bronchial arteries and veins is, that they are the nutrient vessels of the lungs, maintain their tissues, and supply them with blood - life. May we not already say, give them individuality? But with regard to nutrition as a life-function there will be more to remark by and bye.

26. If the whole of the arterial blood of the body in its newness of life from purification and aeration exists in quick successions in the *rete mirabile* of the lung-vesicles, that blood, which under the auspices of the brains nourishes every other organ in the body,

should nourish the lungs themselves *à fortiori* and even *à priori*. Why should it not? It has not indeed passed through the left heart, and received its innervation, and may therefore be called prae-cardiac; yet for the lungs it is arterial blood. A reason against its nourishing the lungs is found in the bronchial arteries, small rills of blood also in the lungs, and which want a function, and the nutrition of these organs is assigned to them. In the embryonic state they not only nourish but also lay down and so to speak construct the lungs, but in that state the tides of the heart are not supplied to the latter. If they ramify through the bronchia, they no doubt are feeders to the lungs, but not structurally or wholly in the presence of the wealth of the finest and prior blood supplied from the inmost by the pulmonary circulation. Their office in this respect to the substance and mass of the lungs becomes external and secondary after birth, as the blood they supply, coming not from the pulmonary depths but from the aorta or one of its arteries, is secondary and external.

27. If then the lungs are bodily supported and nourished by the capillaries of the pulmonary circulation, are there not two ends accomplished, two things brought about, in the fact that two such different orders of vessels are at work in the case? The blood-river of the entire man, body and head, is poured through the pulmonary vessels; a streamlet from the aorta, through the bronchial arteries, goes to the bronchia. Is there a function besides bodily nutrition signalized by this latter income of circulation? Observe that the stress and burden of forming and building the lungs is now taken from these little arteries, that they are inhabitants of these organs, and not now

their architects, and that thus they are perhaps disengaged for new equivalent uses.

The Breathing Lungs and the Beating Heart.

28. Watching from its banks the above river from the heart to the heart, we might imagine that the lungs which seem in the force of it, would be carried thereby into its successive waves of motion; that pulsation would reign, and the heart be all in all. Were this the case, the body would live and not the mind; for the existence of the mind, conscious and unconscious, depends in the lower resort upon the breath of the lungs. Thought, as we may instantly observe, follows it, or rather precedes it, and in waking and in dreaming is equated or corresponds to it. In profound sleep, as in profound thought, the breath meditates

and hovers over the subjacent man. *Anima*, *Animus*, *Spiritus*, the breath of Life, are expressions of these facts.

29. May we seek for the *rationale* of this independent movement of the lungs? Everything is an effect of causes, and causes in nature are reasons to the mind. Reasons, however, do not create things given in creation, but perceive and explain them.

30. In the first place, the brain lives after birth above the expanded lungs, and is a head also over the heart. In the brain there is a spirit which carries the mind, and that spirit breathes, or as Swedenborg says in his psychology, "animates." The lungs do not animate, but are animated by the brains, and so respire. The brain mentalizes the lungs, and therewith the entire body, both sleeping and waking. This would be impossible if the bolt and body of the blood coerced them to pulsate instead of to respire; then there

would be blows and not breaths; not even impulses, but physical impulsions. The respiratory motion environing the pulsatory is necessary before understanding can equate with lung, and descend by influx into it. Are the bronchial arteries and veins the occasion taken by the brain of making room for the lungs in their bill of rights, and making their first breath a declaration of their motory difference from, and independence of, the heart?

Attractive Freedom a Foundation in the Body.

31. It is a physiological and psychological law laid down by Swedenborg, that from the mind, and then from the two brains, and from the lungs, every organ has the power of attracting out of the general circulation the quality and quantity of blood which it

requires for its tissues and functions. So there is no inordinate delivery of any drop or kind of it by the stroke of the heart into the intimate room of any viscus or organ. All is drawn in most specifically by command of the brains and the lungs through the various forms of the organs. Each form breathes the blood in from its circumferences to its centres. The form is determinant of the motion and the stream. So organic attraction reigns, and organic freewill paramount in the attraction. The great common blood of the heart may be likened to a sea whose shores are the boundaries of the organs. In its waves are invisible ships, freighted and chartered, bound each to a port in a special organ, and the cargo expected, and unloaded, for that organ alone. The Captain and Crew of these are from the brains and sympathetic nerves; for their life is in the blood also, and governs in it as well as in the organs. This life in the blood is a compliance with orders reaching from port to port. Furthermore, the ocean and the ships are one, only differentiated by the nearness of the vessels to the harbours: every billow as it draws to its destination is a manned ship for some organ haven. So this blood of man, *mar lidha*, is not *mare navigerum*, but *mare navium*, not " sea with ships on it," but " sea of ships."

32. This law is typically illustrated in the lungs in their discriminated liberties. They receive perforce all the blood of the system, and then pass it through their inquisition, which becomes more and more severe and liberal to the inmost judgments or vesicles. But being the great bodily attractors and tractors of all, their traction and attraction are different in each lobe, lobule, cluster of vesicles, and single vesicle. Each globule of impure blood is therefore scrutinized

individually or atomically at last, and purged into the air from its minutest habits and secrets in the infinitesimal dock that belongs to it. So that though the lungs receive the blood natively from the propulsion of the heart, they take it over at once into their own custody, and command and more and more exactly rule the pulmonizing distribution and doom of it. But being thus at length arterial blood, and passed from night into morning, it is again socialized into little streams, these into larger, and so on until it forms the great disemboguements of the pulmonary veins, which empty themselves into the left auricle, and are great arteries from the lungs.

The Bronchial Vessels before and after Birth.

33. Before birth the fœtal lungs, unopened to the air, were moved by the strokes of the

bronchial arteries as masons surrounding and building them. The whole blood of the heart passed through channels internal to itself. The nervous system and the brain commanded this result. In other words, the animations or life-movements of the two brains consented to, or were consentaneous with, the movements of the heart, and with the inaction of the lungs. After birth these animations consent to, or are consentaneous with, the movements of the lungs. What is the bridge which connects this new state with a new function or use of the bronchial arteries and veins?

34. It lies in the end for which the foresight of the brains governed the bronchial arteries to make the lungs as forms for future respiration. Those arteries nursed and brought up the lungs for the functions that were to arise from their constitutional forms. The lungs were made in order to

breathe,—to respire. The heart was not informed of this, but a messenger from the heart was privily commissioned by the brains to execute it. Now the bronchial vessels are comparatively remote from the heart, and its force is minimized or lessened in their channels. They furnish the lungs with a small peculium of blood which is their own property, and which they can dispose almost in their own way. Here perhaps is the foundation of a freedom which begins breath, which begins the voice of the lungs, which is their constitution of liberty.

35. The air of freedom, the free air, leave to breathe, are words that the lungs have coined in the social, political, moral, and spiritual mints. They are true tokens. No senate disowns them.

36. This mode of supply of matter to spirit is a rule in our form and organization. Everywhere matter, uncapped, and some-

times uncoated, has to go in at a side door, or a kitchen-door, to be admitted to its business of mere necessary clothing and supply. The bronchial arteries come off at angles, which breaks their stroke. The carotid arteries twist and curve, and lay aside their coats, in the rocky spirals that let them to the brain. The aorta by its arch moderates the heart-force for the whole body. Its first great rami, the arteriæ innominatæ, the subclavians, the carotids, and the rest, are all modulations and diversions of the strong current. The lateral branches further down are in the same concession. Each is equated to the freedom of the organ which calls for the blood: every angle and clause of each is a stern prudence and stopcock of supply. The straitest railroad of the arterial blood is for the limbs and voluntary muscles, where the will is in its rights for great demand of heart-force and proper action. Were it unhindered elsewhere, it would govern the inside wheels also, and they would be no better than its arms and feet. But by soul-arrangement they are wheels within wheels, a cherubic protection. The Spirit of the living creature, and the breath under the spirit, are in them.

37. Thus it may begin to seem that this small blood-system of the bronchial arteries and veins gives to the lungs as individual organs distinct from the heart, though closely united to it, a life of their own, in addition to, and in separation from, the manifest life which they draw from the heart. For if they had not a life of their own, they could neither be conjoined as they are to the heart, nor enjoy any individual function. This conclusion is not adverse to the current opinion of physiologists, that these bronchial arteries and veins are the nutrient vessels of the lungs. For this view implies that they

nourish the tissues for the sake of the functions and movements and ends of the lungs, all of which,—namely, attraction and reception of the blood, winnowing of it, purifying it, transmitting it, absorbing special lung-blood from it,—are contained in respiration: of which thus these bronchial vessels are the opportunity or immediate apparent cause. To supply adequate agents with means to a work is to prosecute the work. This nourishment also implies conservation of spirit or identity. The bronchial arteries which give the lungs a little sufficient property of their very own, just enough for them to keep and hold, are the foundation of a suffrage or individuality throughout them which makes of them a multiform free province in the United Kingdom of the body; the contagion of which draws the whole body into freedom. Think what a free suffrage is: a breathing elbow-room of personality, a birth of it: at first an ignorance, a babyism, a nothing, a privilege, a wish, a vote, a voice: but in its lateness and in its gathering a loud commonwealth. We are forced to illustrate these little star-points, to bring them to the megascope, and away from the microscope; and to see that to truer knowledge they are worlds.

38. May we say here *en passant* without offence to a great body of earnest workers, that Physiology at present belongs to the heart, and not to the heart and the lungs conjoined. The strokes of work are incessant, the considerations that should accompany them are fitful and asthmatic. In the infinitesimal field where much of the work lies, there is no breathing life of induction and deduction. For nearly fifty years I have pleaded with my brethren to look to the doctrine of respiration as Swedenborg has taught it, for it is the soul of physiological

physics, and the gate to psychology. It is as easy to be seen, if you will, as breathing is easy. The only difficulty is that it is not near you, or far from you, but it is yourself. But as teaching, as philosophy, it ultimately imports that the movement of respiration involves the performance of all function, and the perpetual union of the will and the understanding in consciousness in the body. It is not irrelevant to this that the primeval gift of Life is represented thus: "And the Lord God formed man, the dust of the ground, and breathed into his nostrils the breath of life; and man became a living soul" (Gen. ii. 7). I shall have to recur to this subject in the following pages. It is sufficient here to recommend my kind readers to study the whole case in Swedenborg's *Animal Kingdom*, where it is written of in letters of light.

The Mission of the Lungs.

39. Thus far we have seen that the bronchial vessels give the lungs the power of working their own form and functions independently in the known rhythm of respiration. We have found that these vessels yield them a peculiar blood not a part of the cardiac-pulmonic circulation; a restrained gift, as it were scattered to them as birds are fed, in handfuls from a distance; so that they, the lungs, can pick it up at leisure, or, in fact, in freedom. They exercise a delicate act of reception of this modicum coming in widely different circumstances from those in which the whole blood pours from the heart through the lungs to the heart. Two nutritions are here: one, of the lungs for their circulatory service in the pulmonic circulation; the other from the bronchial circulation for

their respiration alone. As the former involves the body through the heart, so the latter involves the body through the lungs. For that freedom called breath is communicated by a universal gear of ligaments, membranes, muscles, fasciæ, to the frame, from the top of the head to the tips of the toes, adown the neck, through the breast and through the belly; and pulls out the whole elastic man into expansion, and lets him go into subsidence, with each inspiration and expiration. The bones and every tissue, however unyielding, are drawn as far as they can yield, or tides are made in their fluids. Thus the fomenting of respiratory acts, the nutrition of breathing, gives the whole of the body and the parts a motion of freedom, and so a power of selecting and administering the blood-material which belongs to it, organ by organ, as the nerves ruling in each case require. This amounts to the further position that this freedom is the corporeal analogue of the intelligence which is in the mind, which reigns in the brain, and which representatively mentalizes the lungs, so that if you watch them breathe, they cast ghost-shadows, or seem as doubles to be thinking your thoughts with you.

The Passage to Psychology.

40. We seem now to have said enough according to our insight to open and illustrate the above first doctrine of the Respiration as a universal function in the body, and as excited by, and seconding mental function. This doctrine further considered paves the way to the truth that the whole body is animated with the similes of life which it derives from its brains and from the mind therein ; that it is nothing but an ordinate continuation of mind let down for lower

services; that is, of will and understanding in organic forms, in the chest and in the abdomen; forms in health unconscious for us; but in themselves conscious carriers of mental principles and states. For we know that in a perfectly healthy body each molecule and organic and fluid unit keeps its place, acts its part, and ministers its obedience, better than nations, societies and families and men and women know and do their *duties;* better even than instinctive animals know and manage their lives. The inference from the imperfect head and mind which preside, or do not preside, in each of us, is, that the form and frame of every created man is a mental thing; that what is outside him is used mentally: in other words, that our organism is an agent, efficient through ends, causes and effects, either by its own intelligence, or by an intelligence perpetually communicated and ingiven by Spirit Itself and supreme Intellect Itself. " The light of the Body is the Eye." Ergo, the eye lives all through the body : the body is ocular : every atom sees, or how could it work in our underworld, in the mines of the flesh with no safety-lamp? The divine context proceeds, "If thine eye be single, thy whole body shall be full of light : " representing the spiritual-corporal omnipresence of that mind here called the eye —the omnipresence of its light, or of its darkness. We pursue this subject further in our section on a Doctrine for the Human Body.

41. [The principles which stand at the head and beginning of a series are continued throughout it, but no other principles are so continued. A tree involves ends, causes and effects from the Divine as a man does. But a tree has roots or seeds, not conscious brains. Its structure and perpetual construction imply a creative intellect caring for it, but not a conscious intelligence in itself in any part.

Beginning from an unconscious seed, it is unconscious all through, from seed to seed. Yet it is full of the representation of the Supreme Intelligence. We write this lest it should be inferred that the exhibition of mind in consciousness can exist out of the series of a conscious brain.]

The Will and the Understanding.

42. These preliminaries are suggested by Swedenborg's doctrine of the office performed by the bronchial vessels in the lungs, corresponding as it does to a spiritual doctrine holding on the mind, on the relation of the understanding to the will. It is proper now to give some extracts from his *Divine Love and Wisdom* opening up the high correspondences of the heart and the lungs. From these also we may have better light on the physiological doctrine.

43. The human "Will," observes Swedenborg, "is created and formed by the Lord in man as a receptacle and habitation for His Divine Love, and the Understanding for a receptacle and habitation for His Divine Wisdom. The will and the understanding are in the brains, in the whole and in every part of them, and thence in the body, in the whole and in every part of it. There is a correspondence of the will with the heart, and of the understanding with the lungs. All things of the mind are referrible to the will and the understanding, and all things of the body to the heart and the lungs. By the correspondence of the will and the understanding with the heart and the lungs, there is a correspondence of all things of the mind with all things of the body. Many *arcana* concerning the will and the understanding may be discovered through this correspondence. The conjunction of the spirit with the

body is by the correspondence of the will and the understanding with the heart and the lungs; and disjunction through non-correspondence."

44. "In the world it is scarcely known what the will is and what the love is, since man is not able to love, and by love to will of himself, in the same manner as he is able to understand and to think as of himself. A parallel case is that he is not able from himself to influence the heart to move in the same manner as he is able from himself to influence the lungs to respire. Now, as it is scarcely known what the will and the love are, and it is known from sight and examination what the lungs and the heart are, therefore when it is recognised that these two pairs of things correspond, and by correspondence make one, many *arcana* concerning the will and the understanding may be discovered which cannot be detected otherwise."

45. A spiritual Algebra is for the first time created here. The heart and the lungs are known numbers or quantities; the will and the understanding are the x y z, the unknown reals which are to be equated and illustrated into known things, but on a higher plane of heart and lungs, on which plane they are the organs of the love and wisdom of the man, and for him the receptacles and habitations of the divine life of love and of wisdom.

46. In Swedenborg's formula we have not only the beginning of a spiritual Mathesis, but also of a new height and latitude of inductive spiritual science, in which for the first time great planes of observed fact communicate by a newly given power of thought with a broad upper world, made visible as the process of induction proceeds in reality above answering to reality below. The first ascertainment is that love is the spiritual heart, and that intellect is the spiritual lungs.

What a world of correspondence — even scientific correspondence—now comes within hail!

47. Swedenborg says again: "All the things which can be known of the Will and the Understanding, or of the love and the wisdom; therefore which can be known of the soul of man, may be known from the correspondence of the heart with the will, and of the understanding with the lungs."

48. "Because love and wisdom in the Lord are *distinctly one*, and the divine love is of His divine wisdom, and the divine wisdom is of His divine love, and because these proceed in like manner from God-Man, that is, from the Lord; therefore those two receptacles and dwelling-places in man, the will and the understanding, are so created by the Lord, as to be *distinctly two*, but still to make one in all operation and in all sensation; for the will and the understanding cannot be

separated in these. *But, in order that man may be able to become a receptacle and a dwelling-place, by the necessity of the end the fact has followed, that the intellect of man is able to be elevated above his own love-proper, into a certain light of wisdom in the love of which the man is not, and by this to see and to be taught how he ought to live, to enable him to come also into that love and so enjoy beatitude to eternity."*

49. " After birth the heart immits the blood into the lungs from its right ventricle; and when the blood has passed through them, emits it into its left ventricle; thus opens the lungs. This the heart does through the pulmonary arteries and veins. The lungs have bronchial pipes which ramify, and at length end in air-cells, into which the lungs admit the air, and so respire. Around the bronchia and their ramifications there are also other arteries and veins named the bronchial,

arising from the vena arygos or vena cava, and from the aorta. These arteries and veins are distinct from the pulmonaries. From this it is plain that the blood flows into the lungs by two ways, and flows out of them by two ways. *Hence, it is that the lungs are able to respire non-synchronously with the heart.*"

50. In suffocation and swoons, the heart alone acts and not the lungs, for respiration is abolished for a time. "The blood in the meantime makes the circle through the lungs, but through the pulmonary arteries and veins, not through the bronchial, and it is these last that give the power of breathing."

51. "*The love or will introduces the wisdom or understanding into all the things of its house.* By the house of the love or will is understood the whole man as to all things which belong to his mind; and because these correspond to all things of the body (as

shown above), therefore by this house is also understood the whole man as to all things which belong to his body, all the members, organs and viscera. That the lungs are introduced into all these things in like manner as the understanding is introduced into all things of the mind, may appear from what was shown above, namely, That the love or will prepares a house or bridal bed for the future wife which is the wisdom or understanding (n. 402): and that the love or will prepares all the things in its human form, that is, in its house, to enable it to act conjointly with the wisdom or understanding (n. 403). From the particulars given under these two heads, it is plain that all and singular the things in the whole body are so connected by the ligaments emitted from the ribs, vertebræ, sternum and diaphragm, and from the peritonæum which hangs on these supports, that when the lungs are respiring, all are drawn and borne in like manner into alternate acts. Also anatomy clearly shows that the alternate waves of respiration actually enter into the viscera to their very inmost recesses; for the ligaments above mentioned cleave to and cohere with the sheaths of the viscera, and these enveloping sheaths dive down to their innermost parts by exsertions or prolongations of themselves, as do also the arteries and veins by ramifications. Hence it may appear that the respiration of the lungs is in plenary conjunction with the heart in all and singular the things of the body: and that the conjunction may be exact and complete, even the heart itself is in the pulmonic motion; for it lies in the bosom of the lungs, and it cleaves to them by the auricles, and reclines upon the diaphragm, by virtue of which its arteries again participate in the pulmonic motion.

Moreover, the stomach is in the like conjunction by the connective coherence of its œsophagus with the trachæa. These anatomical facts are adduced in order that the nature of the conjunction of the love or will with the wisdom or understanding may be seen ; also of the conjunction of both in consort with all things of the mind : for the anatomical conjunction is the simile of the mental." (*D. L. W.* n. 408.)

52. " These statements may be seen in an image, and be confirmed, by the correspondence of the heart with the love and of the lungs with the understanding, concerning which see above ; for if the heart corresponds to the love, then its determinations, which are the arteries and veins, correspond to the affections, and in the lungs to the affections of truth : and because in the lungs there are also other vessels, namely the air-vessels, through which the respiration is carried on,

therefore these vessels correspond to the perceptions. It is always to be remembered that the arteries and veins in the lungs are not affections, and that the respirations are not perceptions and thoughts, but that they are correspondences, for they act correspondently or synchronously. In like manner it is to be observed of the heart and the lungs, that they are not the love and the understanding, but that they are correspondences; and since they are correspondences, the one thing may be seen in the other. Any one who is thoroughly acquainted with the anatomical fabric of the lungs, and collates it with the understanding, may see clearly that the understanding does nothing of itself, does not perceive or think from itself, but does all from the affections which are of the love; which in the understanding are the affection of knowing, of understanding, and of seeing the thing; which affections were treated of above. For all the states of the lungs depend on the blood from the heart, and from the vena cava and aorta; and the respirations which are performed in the bronchial ramifications, exist according to the state of these blood-vessels; for if the influx of the blood ceases, the respiration ceases. Much more may still be discovered by collating the fabric of the lungs with the understanding to which it corresponds; but as anatomical science is known to few, and demonstrations or confirmations by things unknown place a subject in obscurity, therefore it is not well to say more on this theme. My knowledge of the fabric of the lungs fully convinced me, that the love through its affection conjoins itself to the understanding, and that the understanding does not conjoin itself to any affection of the love, but that it is reciprocally conjoined by the love, to the end that the love may have sensitive life and active life. But it is to be constantly borne in mind that man has a twofold respiration; one the respiration of the spirit, and the other of the body; and that the respiration of the spirit depends on the fibres from the brains, and the respiration of the body on the blood-vessels from the heart, and from the vena cava and aorta. Moreover, it is evident that thought produces respiration, and it is also evident that the affection which is of the love produces thought: for thought apart from affection is just like respiration without a heart, which is not a possible thing. Hence it is plain that the affection which is of the love conjoins itself to the thought which is of the understanding, as was said above, in like manner as the heart conjoins itself [to the respiration] in the lungs." (*Ibid.* n. 412.)

53. "*The wisdom or understanding, by virtue of the potency given to it by the love,*

is able to be elevated, and to receive those things which are of light from heaven, and to perceive them. That man is able to perceive the arcana of wisdom when he hears them, is shown above in many places. This is the faculty of rationality which every man has by creation. Through this, which is the faculty of understanding things more internally, and of deciding what is just and right and what is good and true, man is distinguished from the beasts. This therefore is what is meant by the understanding being able to be elevated, and to receive those things which are of light from heaven, and to perceive them. That such is the case, may also be seen in an image in the lungs, because the lungs correspond to the understanding. It may be seen in the lungs by the cellular substance, which consists of the bronchia continued down to the minutest follicles, which are the receptions of the air in the respirations: these are what the thoughts are one with by correspondence. This follicular substance is so natured that it is able to be expanded and contracted in twofold states; in one state with the heart, and in the other state almost separate from the heart. In the state together with the heart it is expanded and contracted through the pulmonary arteries and veins, which are from the heart alone; in the state almost separate from the heart, through the bronchial arteries and veins, which are from the vena cava and the aorta: these last vessels are external to the heart. This is the case in the lungs, because the understanding is able to be raised above the love-proprium which corresponds to the heart, and to receive light from heaven: but still while the understanding is raised above the love-proprium, it does not recede from it, but draws out of it that which is called the affection of knowing and understanding, with a view to somewhat of honour, glory or gain in the world : this somewhat cleaves to every love as a surface, and on the surface the love is lucent from it; but with the wise it is translucent. These things are brought forward about the lungs, to confirm the position that the understanding is able to be elevated, and to receive and perceive those things which are of the light of heaven ; for the correspondence is plenary. Out of that correspondence it is easy to see the lungs from the understanding, and the understanding from the lungs, and thus to have confirmation out of both at once." (*Ibid.* n. 413.)

54. " This conjunction and disjunction of wisdom and love may be seen as it were effigied in the conjunction of the lungs with the heart. For the heart is able to be conjoined to the clustered vesicles of the

bronchia by the blood emitted from itself, and it is also able to be conjoined to them by the blood emitted not from itself but from the vena cava and the aorta. Hereby the respiration of the body is able to be separated from the respiration of the spirit. But when only the blood from the heart acts, then the respirations cannot be separated. Now, because the thoughts by correspondence make one with the respirations, it is also plain from the double state of the lungs in respiration, that a man is able to think in one way, and of his thought to speak and act in one way, in company with others; and to think differently, and of his thought to speak and act differently, when he is not in company, that is, when he is in no fear of loss of reputation. For he can then think and speak against God, the neighbour, the spiritual things of the Church, and against moral and civil laws: and he can also act against them, by stealing, by taking revenge, by blaspheming, by adultery. But in companies where he is afraid of losing his character, he is able to speak, preach and act just like a spiritual, moral and civil man. From these things it may be seen that the love or will, in like manner as the understanding, is able to be elevated, and to receive those things which are of the heat or love of heaven, if only it loves the wisdom in that degree; and that if it does not love the wisdom, it can as it were be separated." (*Ibid.* n. 415.)

55. " Now, because the love corresponds to the heart, and the understanding to the lungs, the aforesaid positions may be confirmed through their correspondence : as for instance, how the understanding is able to be elevated above the love-proprium into wisdom ; and how the understanding is dragged back from the elevation by that love if it is merely natural. Man has a

twofold respiration, one respiration of the body and another of the spirit. These two respirations are able to be separated and also to be conjoined. With merely natural men, especially with hypocrites, they are separated, but rarely with spiritual and sincere men. Wherefore the merely natural man and the hypocrite, in whom the understanding is elevated, and in whose memory therefore many things which are of wisdom remain, can talk wisely in company out of thought from the memory; but when he is not in company he does not think from the memory, but from his spirit, thus from his love. He also respires in like manner, since the thought and the respiration act correspondently. That the fabric of the lungs is such, that they are able to respire by virtue of blood from the heart, and by virtue of blood from outside the heart, was shown above." (*Ibid.* n. 417.)

56. It may be well to remark here, that when Swedenborg speaks of gaining knowledge of the will and the understanding, and their reciprocal action, conjunction and relations, from the anatomical structure and physiological functions of the heart and lungs, he in no way sets aside our ordinary consciousness, or ignores the common sense and conscience of mankind, as if these were not plain guides in the path of life. His works are written for all who can read them, and many at present cannot; but those works especially which treat of a new divine philosophy, are for those who need that particular bread of life; whose bent it is to investigate subjects in their depths, and who suffer disappointment and states of doubt if they cannot have satisfying light on why? and wherefore?—on ends, causes and effects. They are the poor in faith, and not often the poor in spirit, and a gospel of philosophy is

E

for the first time a table prepared for them in the presence of their enemies, negation and its agnosticism.

57. From this new departure we may foresee, that what has hitherto been metaphysics will become tangible and organic when the human frame, which is the ultimatum and vessel of all internal speculation,—for no embodied man ever yet had a disembodied thought or liking,—comes to stand in its rationated organism, as a set of known quantities over against the old invisibles, will, affection, understanding, perception, thought, imagination, and the like; and to summon them through order into their incarnation in their organic seats. Wherever such philosophy is valid, and has any real issue, this will be demonstrated by structural thought, "travailing into the human form." Of such structural thought Swedenborg's works are the consummate exemplar, and

this "kingdom of man" comes to knowledge through them. Poetry indeed has had structural imagination, and has been eminent in side issues and lights accordingly. But Poetry of itself has no firm walk of structural thought, but in its "fine frenzy," is always at its best in momentaneous apparitions, which cannot continue, but presently fade away. We remarked above that a new platform of mind is needed for these new results. The dry formula that the lungs correspond to the intellect, and the heart to the will, launches into humane knowledge a power and method of induction hitherto unattainable; and the length and the breadth and the height of it are equal. It runs from natural planes to greater spiritual planes, and increases in substantiality and its light as it ascends. The first Seer of this was soon consciously in the Spiritual World.

The Mental Side of the Correspondence.

58. "Such is the correspondence between the will and the heart, and between the understanding and the lungs, that what the love does with the understanding spiritually, that the heart does with the lungs naturally." It is the longest and broadest fact in human and animal nature, that what the heart most desires,—meaning here by the heart the ruling love,—it strives by all ways to carry out through the assistance of the mind. This holds of all wants and wishes which clothed with thoughts soon become hopes and aspirations,—from the gratifications of the senses, to the love of wealth, of power, and of empire. In the beginning of life this strong current is well-nigh irresistible, and often remains so in the adult man. It is checked by instruction, by education, by fear

of consequences, by experience of disasters; either by moral and religious, or by worldly and selfish considerations. It lives as a fundamental bias and genus even in a regenerating man : it is whole and sole life, hope and struggle in a man abandoned to himself and the world.

59. This first state for all is the first conjunction of the mental heart with the mental lungs. The man accepts his nature and existing character as " very good." Apart from other motive than the huge good pleasure of these, he would rush into action as though the neighbour and the world were airy nothings, or playgrounds for his will. But mercifully to the heart-maniac external motives do come. Rubs come. Instruction comes. Always it must be an instruction suited to the man to be of proper use. In other words, it must have a hook of reference to his character and disposition, but a slender

hook. It must be no direct artery of the heart and body of his passions, lusts or desires; yet, being partly of their nature, it must win by insinuating that their safe existence and gratification lies in the way of order and restraint. In short, disallowing their forces and pretensions as rulers and masters, it must win them to hearken to itself as a derivation emissary and comrade from themselves; and in other words, to pause, take breath, and consider the situation.

60. The tyrannous rush of blood from the heart through the lungs to the heart is the unchecked dominion of Will. The small gentle trickle and thrill through the bronchial arteries, is the will also, but accommodated to yield to a regime of instructions, informations and observations.

61. The mind can exist on these terms, but not on the heart-alone terms: a commencement and increase of understanding

can exist. Bear in mind that the intellect has not to be assumed, or brought in for a theory, as *deus ex machina;* but is there, ready to countervail the tides of the will, and has gentle considerations with it, so soon as it perceives that it has a better self of its little own; still a self, still a self-will; but a manageable one; a pony for the child, a horse for the man; instead of a lion or a tiger. The man's nature now then can breathe. The pulmonary arteries and veins, the heart-system, has submitted somewhat, and has caught touch of the diverse life receptive of the respiratory movements, and acting on and with the bronchial and now the lung-system.

62. Watching the mind, the conscience, the common consciousness, we observe this every moment. Rational action, self-management, guidance, and obedience to right, is nothing else than this. At first we perceive

a benefit in doing well, a true perception, and act upon it. Here is a twig of the bronchial artery feeding an air-cell or thought. Then the air-cell becomes a perceptive thought, draws the artery or affection more to its own way of thinking. Each thought breathes of necessity to a new thought, or it ends in a drowse. Understanding, the mind's sight, is in this way introduced and established, as correspondentially respiration is established. The distinct affection which supports the understanding, — the bronchial blood which feeds the air-cells,—is the agent throughout.

63. The understanding still anchored to self, seeking rewards of benefit,—pleasure, honour, glory, or gain,—and again pleading "very good," is open to new instructions, from parents, masters, from the Word of God, and begins to rule down the self of itself so as to form a head and a body to it: out of chaos and protoplasm to make

a child of it, and the head comes to the top, and the stomach to the middle, and the feet to the ground, and the man becomes muscular and erect. Himself, by might of heaven, has travailed into the human form. This is the name of a new will and a new understanding. Every will is aggressive, righteously or unrighteously. The new will in which obedience is wrought, dominates without destroying all other body, parts and passions in the man. In like manner the respiration holds the heart round its neck in its embracing arms, and the lungs in their depths now dominate the circulation, and for service and use possess all its wealth. So at first the bronchial vessels correspond to secondary selfish considerations causing the first breaths, and lend life to these, but at length correspond to ministration at the footstool of the throne of reasons supreme over the actions of the will.

The Separation of the Intellect from the Will.

64. Synchronism between two organs means that they can keep time with each other like two even-going clocks in a house. Movement with absence of synchronism, with marked irrelevance to it, implies incommensurateness in times and rhymes. Now time in the body corresponds to *state* in the mind. Therefore want of synchronism between two organs of the mind means diversity of states of activity between them. Such diversity exists between the acts of the will and the intellect. A man's will, the maniple of all his wishes, the propellent power of them, is one thing. His intellect, taking cognizance, more or less discriminating, of these claimants to volition, confronts them, blocks their way, and shows their quality, possibilities and

probabilities to the man. Are they able to be carried out on the lines of good perception and true thought, and of conscience? Are they feasible to be done in the face of the world? Are they good policy for a respectable man wishing to advance himself? Will they minister to success for himself and his house safely enough, or long enough, to be worth carrying out supposing they are wrong? Will they add peace to the mind at last? The intellect or understanding—we use the two terms indifferently—imports these considerations, and in this its state and office is undoubtedly separate from the will. Of itself, will knows nothing of such checks and motives of pause: it feels them from its admonishing mate, the intellect. The two are separated in order that in all life perpetual schoolmaster and perpetual pupil may be united, endowed, and blest in a spiritual man. If the will and the intellect were identified instead of active and reactive, if the thoughts of the mind in its spirit or breath were the same with the intentions of the will in its impulses or beats, and if the pulses ruled as they must do, the man, if he still lived, would be improperly born, an abortion wrecked upon nature. The first cry to get breath precludes the monster. Yet the man afterwards can approximate to him, and come near him, though not quite attain him, by giving free rein to his lusts and passions, in disregard of the voice of conscience and reason.

Endowments from the Separation.

65. These endowments are both good and evil. For all men, in the fair justice and judgment of the Lord of Men, there is indeed an intellect which *will* speak, which *will* dictate right and wrong, and *will* proclaim

consequences. But the intellect can be corrupted by the will, cease to admonish it, and be perverted by it, and inverted. To the worldly and selfish mind in predominance, such an intellect dictates prudence in carrying out its designs, and helps them by its cunning. Here it is a high watch-tower from which the mind surveys the field of its intrigues with all its optics and instruments. It can develop superb talents in the man, as becomes the intellect as king of the emulous passions. It can soar in keen perceptions, outwitting even clergy, and elevate itself to all religiosity, even to the pinnacle of the Jewish temple, and quote scripture for the top.

66. This common use of the intellect corresponds to the rule of the heart in the lungs by the blood alone; the respiration serving to confirm the rule, and to carry it out into a systematic character. The pulmonary vessels are chief; the vesicles of

the lungs, the first thoughts of the man, are the *prima falsa;* the bronchial system is the stamp and signature of the state : the courtier and hypocrite of the royal pretexts of the will. It still shows capacity as breathing-in the great world : it keeps the body and the blood from unmannerly apoplexy, as the man from maniacal perdition here. It compels him still to breathe; and therefore to think; to scheme : and may be sometimes to consider.

67. To the regenerating man, to him who fears God and would fain love the Lord and the neighbour, who recognises from these grounds right and wrong, and strives to live the truths revealed by the Lord in the Word, the intellect is separated from the will so far and so heartily, that from the pinnacle of the Christian temple it sees the will in the depth of evil and corruption, and with "Get thee behind me, Satan," rebukes its advances.

Then more and more shielding itself from the actual will, and severing itself from the hereditary will, this intellect brings forth new thoughts and truths from the new life's affection; and these, when once acted upon, become the fibres of a new will, with the sure accompaniment of an added and previously inconceivable understanding. So the heavenly marriage of the will and the understanding, of the spiritual heart with the spiritual lungs, summons a train of angelic presences and ministers to the man, revealed to him only by perpetual new loves and vernal comprehensions.

Further Bronchial and Pulmonary Considerations.

68. With respect to this latter use of the intellect, it is identical with the respiration of the spirit or the higher will and under-

standing flowing down into the lower, and governing it; and then descending into the body, where they find their basis in a correspondent organism. As we have already seen, the respiration drawing upon the bronchial vessels for its blood of life, represents that provision. It amounts to a new will in the will conjoined to a perpetually renewed or ever-breathing intellect. The will is the whole man, and the new will so governs that the old will is bodily subject to it, and under it. The will and the intellect elevated equally are then at one. This perhaps attains to something of what Swedenborg enjoins where he says,—" Examine the connexions well, and survey them with an anatomical eye, and afterwards, following these connexions, consider their coöperation with the breathing lungs and with the heart; and then instead of the lungs think of the understanding, and instead of the heart think of the will, and you will see." It is a deep and difficult subject excepting in its simplicity and to the simple. In its interiors it is full of arcane treasures to the spiritually-intelligent man. Swedenborg has manifested it for those who can understand, or, as he says, "see." But the one point which he has forborne to press further lest he should confuse the reader with anatomical details, is the office of the bronchial vessels in making possible for the body a respiration alien in its times to the pulsations of the heart: this condition being the platform in the representative bodily man of the separation of the intellect from the will in the real man. Swedenborg counsels the anatomical reader to work and solve the problem for himself; examining what parallel to the bronchial arteries in the body can be found in the affections of the mind; for the parallel found will give to knowledge the bronchial

F

arteries of the mind. Our Shakespear speaks of "the mind's eye." The Lord says, "If thine eye be single;" meaning the eye of the mind. If the mind has an eye, it has the human form, and therefore bronchial arteries. The spiritual *uses* of the organs, always embodied in the forms corresponding, and in the activities of the forms, are the substantial human body in the spiritual world.

A Question from the Spiritual Side.

69. If the native will of man be corrupt, how, it may be asked, can the understanding influence it to good? The instruction of a wholly corrupt will could only, can only, deepen its corruption, and make it more powerful in evil. A practical answer is that man, from his intelligent mind, divinely-given, through infancy when evil does not appear, and through all after-states of tenderness;

through the survivals and remains of these childhoods whenever lived ; and then through primary and lifelong instruction, has a power imparted to him by a Being stronger than will, of resisting his hereditary and actual nature : and then in the struggle and agony of the combat, a new will called conscience is heaven-born through and *in his intellect*, of which it is the supreme form. This is so. It is a miracle, and the divinity performs it. But though miraculous and beyond nature, it is a fact of experience; and like all true miracles, absolutely accounted for on the simplest of principles, that it is a needful intervention where the creature is otherwise helpless; and that the Lord's mercy wills the salvation of all, and has provided adequate means to this end for every man who chooses.

70. Oxygen and Hydrogen, and gold and granite, and the whole field of natural sub-

stances, are difficult to account for compared to the supreme self-evidence of want and supply in such daily and hourly cases. For God is nearer than near to us.

More about the new Will in the two Organisms.

71. Now, when once the new will in its small brook and streamlet,—in its first alliance with the bronchial arteries,—has been accepted into the mind and born into the thought; that is, into the breathing, hope, and aspiration towards it; a position is taken up in the blood of the flesh which commands. A baby king is born. A king who in adult time can be all compact of royal thoughts, and live and breathe for the sole dominion of good. To him henceforth there is no equipoise in the mob and demagogy of the passions; themselves now ceaselessly throb-

bing and pouring forth as of old, but towards a shore of compression and consideration: towards rocky steeps of self-denial. The huge moving masses will spray and scatter in long details of time. The first step is conversion. An atom of it, one resolute thought, one prayer, can turn the tide of the sea of blood, and rule its heart as with an awful trident. It is a process gradual and eternal for men and for angels. The radical self still remains, a quiescent foundation, an acknowledged idiosyncrasy; a "wallet" for "oblivions;" also, with strong reminders, a place of rejection for hindrances to regeneration here, and to progress hereafter.

72. Self-preservation in the highest sense, and its motives, are imported in the separation of the intellect from the will. For the affections clothed with thought in the understanding become visible motives :—instructed motives for avoiding evil and its consequences,

and for seeking good. These motives spring from different parts of the mind's affections, and often at first, as in infancy and in childhood, are personal and selfish. We abstain from naughtiness and disobedience because some punishment has followed. The same considerations avail in after-life. In them we fear God, and so obey Him. Or the like motives of fear for loss of good name, discipline us, and maintain us in society. On the other hand, a holy fear is possible in which there is a love of God, and with it a reverence for our own minds as His work, and as recipients of His mercies. With this comes a fear of hurting or destroying our souls. These states belong to intelligent perception and thought more or less apart from congenital nature, and betoken different stages of the separation of the intellect from the will, and of the government of the will through the intellect. We ask tentatively whether they are not represented in the different origins of the bronchial arteries in different individuals; for the origins of these are various, though they all seem to be from arteries supplying blood-life to the respiration. Generally they come from the aorta, sometimes from the intercostal vessels, sometimes from the œsophageal? Does this import representatively the various first motives of thought for the good of the mind? Where they come off immediately from the great aortic blood, does this imply more egotistic and lower motives; and when they arise from collateral and smaller origins, especially from the intercostals, the arteries of lung-movement, does this mean the representation of nobler thoughts coming askance in mystic suggestions? Were such the case, it would open psychology into physiology in new elucidations, and make the arterial system a face and mirror reflecting and expressing the foundations and fundamental relations of human character. It would also show that the lines for regeneration are drawn in the very creation of the blood and the breath; that they outline a tree of mortal life with an immortal life in the sap of it. The argument is from design, with the special design exhibited.

Some Terms explained.

73. The word Love in these pages is not limited to the sexual or marriage sense, or to any outward relations. In Swedenborg's writings it means the life of man and of mankind. The first proposition of his book is, "The love is the life of the man." To discern this use of the term, carry it a step further, and identify it with the Will. "What a man loves, that he wills." Take the love away, and there is an end of the will, and of

the man himself. The will is the love in its resolution to do and to have what it loves or likes. In both worlds it is the entire man, and the whole human form is its form. If it be happy, its ultimate partner is wisdom, which united with it is the man and his truly human form. Also, whatever the man loves becomes finally his good, and his idea of good. This good is again himself, his body and his flesh. We repeat in this series what has occurred before. The love or will or good or *summum bonum* corresponds to the heart. This love, this will, gives forth derivations all through the mind,—specialized and particularized and singularized and individualized loves, throughout its body, parts and passions. These derivations are the various affections, general and particular, universal and singular. They end in a sea of affections which are all the senses, a capillary sea in the human form. When we have an affection for anything, we love it by aiming at it. We love painting, sculpture, poetry; that is, we affect them. We have an affection for the neighbour, and feel it good to serve him; we minister to him in charity, which is love in good works; we love him. These varieties of love are indefinitely many, but they all exist in the human form, and not out of it. They are inconceivable apart from it. No human affection ever outlies a man. As the heart corresponds to the love or will, so the affections correspond to the arteries from the heart, which give structure and life to the body, as the affections give structure and life to the mind. The affections are of all depths, intimate, interior, and more and more external.

74. The most transient affections are the emotions. They are the arrests of the love when the current overflows, or bursts through its banks; and they range through all the

states of actual life. They are overmastering joys and sorrows, gratitudes, melting sympathies, sudden hatreds; fears, horrors, terrors, panics, despairs : agonies : also ecstasies : supreme moments all. Sometimes they are initial points, or turning points, in a career. They can make the heart swell, so that it has to shout through the mouth : they can make it stand into grandeur so that the whole man is erect and gigantic with it, and the face transfigured with light as on a mount. They can make the hair white as though it were a part of the heart, and went with the blank cheeks. Oftener the emotions are the mere flushes and blushes of the outside man. In some persons they seem to exist as internal lakes of feeling which so easily overflow that they are the main facts of the character. They all belong to the heart, but where and when it is least fortified by the ribs of the regulative and equanimous lungs. Where they are most controlled, the emotion becomes more internal; and so they may be deeply stored like the splenic blood.

75. Following this psychology, every love involves its own delight, from blessedness in the higher man to pleasure in the lower man. The delight is the love itself in its sails of success. What a man loves he delights in, and what he delights in he loves. Whatever it be it is "good" to him, but not always good for him. You may know what a man is by what he delights in. All delights are of the heart, the love, the will, the affections, and no delight is of the mind separately unless the heart be in it. Delight in the most abstract pursuit involves the heart as certainly as delight in power and place, in society, in conviviality, or in wife and children. Strike out delight from any mind, and its machine and machination stops. On

the treadmill itself, the pleasure of looking forward to the hour when the grind is done keeps the will in the feet to their monotonous routine.

76. The intellect or understanding also is not to be limited to outward relations, or to the account of the operations of the man in consciousness. Rather and more than this it is the form, power, controller, and glass of the love. It holds the love in its arms as the lungs hold the heart, guiding, directing and admonishing it; pointing out to it the ways of success and failure, and showing it its own face in wisdom or folly. Of itself the intellect has no desires, but love, affection, behind it, within it, is the desiderator. The love at first essays to win the passionless understanding to its side by uniting with it in hopes; and if these are ill hopes, and the love succeeds, the intellect loses character in them. It is an Ariel, a thing of mere air,

until it unites with the heart in one purpose for good or for evil.

77. These explanations seem needed for those readers who are not acquainted with Swedenborg's works, and especially for any such as are involved in the old metaphysics, which, like the mythologies inevitable at the time, have yet no face, head, neck, heart, lungs, belly or limbs to them, because they have no constructive reference to the human form. That form is given and shown to us first here in the body, which is the literal sense of the soul. Metaphysics irrespective of it are now an anachronism.

The Doctrine of the Human Body.

78. "The mind of a man is his spirit, and the spirit is a man, because all things of the will and the understanding of the man are meant by the mind; and these things are in

principles in the brains, and in principiates in the body : therefore they are all things of the man regarded in their forms. And thus the mind, that is, the will and the understanding, commands the body and all its parts at pleasure. Does not the body act out whatever the mind thinks and wills ? The mind erects the ear to hear, and aims the eye to see ; the mind moves the tongue and the lips to speak ; it sets the hands and fingers to do whatever it pleases ; and the feet to walk whither it will. Is the body anything else then but the obedience of its mind ? Can the body be this unless the mind in its principiates be in the body ?" (*D. L. W.* n. 387.)

79. The body therefore consists, not of principiates *from* the mind, but of the principiates *of* the mind. The principiates in this sense are the principles themselves prolonged into their own ultimates or last results and forms.

80. "Thus the mind is the man himself. The first weft of the human form, that is to say, the human form itself, with all and singular the things of it, comes out of the principles continued from the brain through the nerves. This form it is into which the man comes after death, and which is then called a spirit and an angel, and which is in all perfection a man, but a spiritual man. The material form which is added and super-induced in the world, is not a human form of itself, but by virtue of that other form. It is added and superinduced in order that the man may be able to perform uses in the natural world; and also to draw with him out of the purer substances of the world, a certain fixed continent of spiritual things; and so to continue and perpetuate life. It is a point of angelic wisdom that the mind of man not only in general, but also in every particular, is in perpetual travail into the

human form, because God is a Man." (*Ibid.* n. 388.)

81. Note here that the real man is continually building himself or his character by his mind, brain and nerves, and that this is an immortal formation of his whole person, such as it will be seen to be when he dies. The bodily nutrition covers this in with material from the world, but this is the mundane human appearance, but not otherwise the human form. It falls away as scaffolding, to leave the real form visible.

82. A distinction is also made here between the body as the very *obsequium* or compliance or obedience of the mind, and the body as a separate thing obeying the dictates of the mind. The latter position implies that there are two things, the body and the mind ; the former, that the body is the mind produced into a series of ultimates of itself in which it lives as mind, as organic will

and understanding, or as one identical man. We dwelt cursorily upon this in a former paragraph. The difference between the two views is complete. In the first case, the body has nothing to do with the mind but to stand under it, excepting perhaps on the outside, when through the muscles it obeys the volitions of the will. It is here especially that Swedenborg has illustrated the continuity of the mind through the corporeity of the man. On the ordinary view, in the interiors of the frame, in nutrition, secretion, circulation, and the actions called involuntary, the body is exempt from mentality, it vegetates, developing like a tree; and hence, in this regard, the sympathetic and visceral nerves are called the "vegetative" system. Swedenborg's doctrine is, that all this incarnation and embowelment is a covert theatre of will and intellect; of actions and passions correspondent in an interior providential form to the life and volitions and thoughts of the man: that his interiors are in their places forms and ministers of his character.

83. We have a first evidence of this doctrine in what is recognised as the influence of the mind upon the body. The mind makes not the muscles only, but the organs of the body, tired or vigorous, according to its state. Not voluntarily, but involuntarily as it seems, it acts upon the brain first, and then upon the heart and the lungs, and upon the liver and the kidneys, and lower down still. It can make the whole man feel well, or ill; contented, or deprest. If he is bodily ill, disordered or diseased in any organ, but of a good courage and even soul, that mind limits the mischief down to the organ, and keeps it from ascending in vapours of fear and apprehension. So, though the organ be ill, the man

will not be ill. The mind can abolish the impossibilities born of self-indulgence and its sloth, and chase them from the active hands and feet of duty and need. Or if it so please and so love, it can shroud itself in a coffin of habits, and plead the pity and sickness of every part of the body for the ministration of other hands to itself. It can localize sloth, and make organs sick, expecting them into disease, and depositing inertia from the brain into the maundering blood. It can make mortal diseases mortal before their time, and summon them from the graves of heredity. The first action is on the brain: that once debauched, and mentally if not bodily softened, the rest follows. We speak of the evil deeds of the mind rather than of the good, because they are more common or at least more patent.

84. Passions and emotions are another chapter of the same *influx*, for it is influx

and not mere influence. In the embodied man every communication is a real transference of his spirits. By the above states the mind strikes correspondent organs with force, and can assail, arrest, or pervert their functions. It has been recognised since the earliest ages that the head has the conscious mind within it; that the breast feels affections and emotions; that the heart is love and joy and sorrow, and that the breath is of hope and straitness, aspiration and despair. Also it is written upon experience that the liver is of much mettle, and sympathizes in function with anger and its gall : that it uses Cupid's arrows : as the old saw is, *cogit amare*. The moderns find it a great sugar-maker and starch-modifier, as it ought to be, corresponding to cogent courtesies and inclines. Depressing passions, glooms, unseasoned pessimism may come of it when it does not purge its bitter mind. The Spleen, besides much else, is a holiday and playground for the abdominal blood, in which little is asked of it beyond exercitation : it is let out in batches as it were schoolboys, to shake its sides. " Splen ridere facit, cogit amare hepar." " The spleen sets laughter afoot, the liver compels to love." Shakespear, the Archimage of traditions, notes this of the liver :

> "*Prospero.* .... Do not give dalliance
> Too much the rein : the strongest oaths are straw
> To the fire i' the blood.
> *Ferdinand.* I warrant you, sir ;
> The white-cold virgin snow upon my heart
> Abates the ardour of my liver."[1]

The bodily functions in their words have also become mental words, used directly for conditions of mind. Jealousy is yellow,

---

[1] The liver, not to my knowledge mentioned in the translation of Scripture, occurs frequently in the Assyrian and Babylonian poems and incantations. See Professor Sayce, *The Hibbert Lectures*.

is jaundice, is liver: there is also black jaundice and green jaundice : always yellow still. Spleen or explosion of petty wrath is no exception, for here the clotted spleen is meant; the playground blocked away, where the reverse laughter is shaking ague; in the organ, ague-cake. The reins, the kidneys reprove and admonish in the night-season, and care is near to the begetting loins. They are eliminating organs, often especially active during sleep, purifying their blood for the morning. The heart and the reins are used together in the Scripture as spiritual organs. The bowels yearn with mercy in the merciful, and he who is without mercy is said to have no bowels. The stomach, which seems to include both the stomach for the food and the whole abdomen, is courage mentally. It requires no small habit of courage to put into it what it has to stomach. It is mostly treated as a brave stomach. We read of one who "greatly daring, dined." His reliance on his stomach is meant. Good digestion fills the edile man and builds him. It prepares him for anything: for love and for war — Waterloo, — but after breakfast. The stomach has much mental language; especially with regard to hard words, or unwelcome tidings and indigestible proposals: these we cannot stomach. The mettle of horses when beyond easy control makes grooms call them "stomachful." And again Shakespear says :

> "If you dare fight to-day, come to the field :
> If not, when you have stomachs."
> *Julius Cæsar*, Act v. Scene i.

85. So the brain, the heart, the liver, the spleen, the stomach, the bowels and the kidneys are all used in a tradition immemorial for mental organs, substances, states,

qualifications. The lungs are not so used, except by Swedenborg, but the breath or spirit, which testifies to a perception coeval with language that they are neutral or intellectual organs requiring to be united to the flesh or will of the heart in order to be placed. They are the wings of desire, and fly with our wishes towards effect.

86. Seeing all this, it is not much further to see that all these real loves, affections, passions, imaginations, fancies, senses, married to the organs, and fruitful with them in the world's languages, and continually agent and effective there as mental added to physiological forces, present also and intervening upon occasion as they are in the great lives of the body itself,—are themselves the correspondences and superior actors of whatever goes on in the secret councils and gymnasia of the constitution. They are different for every man because he is his own life from

inmost to outmost. And evermore from a mere piece of given nature which his first wonderful body is,—from the block out of the quarry of natural existence,—he is becoming himself. So is he a mighty yet a least society of atoms or individual men, all transacting the one business of his ruling love, life, or *proprium*, in countless avocations. To shape, make and finish him in obedience to his freely chosen ends is the use of them all. Whatever does not minister to this effect is cast out in his final day.

87. Look then into physiology through the glass of this psychology. The theatre of the human form, and more remotely of the human body, from end to end is the kingdom of the spirit, that is, of the will and the understanding. The will, the love, is the all, which united with wisdom rules atomically as well as personally by divine right. Moreover, as God is a Man, and there is no God

for us but the Personal God, so there is nothing intelligible in and for the finite man except the Personal Man.

88. We may also infer from this transfluence and translucence from above downwards, that the kingdom does not stop with our summit ; that the will and understanding themselves are permeable upwards, though constituted in and of a freedom that makes them individual ; thus that all the voluntary parts from the brain-mind to the senses, are penetrated with the contents, portents and consequences of another life to which they are bound. They are voluntary in the felt sense of being voluntary,—a sense divinely given for ever,—and they are everlastingly responsible : but moreover, they are fluxions of the upper and lower worlds. And of them in this light we know and see no more necessarily than we see and know of the heart and its system, or of our other abstruse

parts. Only we do know that He is human, that we are human, and that all belonging to us is human. Our inevitable ignorance of whatever is above us has brought many revelations, that from the Lord we may begin and increase to know the naturally unknowable. For it is impossible to conceive that God will not teach us that which elevates us to Himself, but which we cannot teach ourselves without His aid.

Old-World Psychology.

89. The early English Poet, John Gower, repeats the ancient tradition of the inhabitation of minds in the Organs. He says:

> " The Splen is to Maléncolý
> Assignéd for herbérgerý.[1]
> The moisté Fleumé [2] with the colde
> Hath in the Lungés for his holde

---

[1] *herbérgerý*, place of lodging.   [2] *Fleumé*, flem.

THE LUNGS.

Ordeinéd him a propre stede
To dwellé there as he is bede.[1]
To the sanguine complexión
Natúre of his inspectión
A propre hous hath in the Liver
For his dwellíngé made deliver.[2]
The drié Coler with his hete
By way of kinde[3] his propre sete
Hath in the Gallé, where he dwelleth,
So as the philosóphre telleth.
Now over this is for to wite,[4]
As it is in physiqué write
Of Liver, of Lunge, of Galle, of Splen,
They all unto the herté ben
Servaúnts, and each in his office
Entendeth to don him service,
As he which is chefe lord above.
The Liver maketh him for to love,
The Lungé giveth him wey of speche.
The Gallé servéth to do wreche.[5]
The Splen doth him to laugh and play
When all unclennesse is away.
Lo thus hath each of hem[6] his dede
To susteignen hem and feed.

[1] *bede*, bid.
[2] *deliver*, free, supple.
[3] *kinde*, nature.
[4] *wite*, observe.
[5] *wreche*, revenge.
[6] *hem*, them.

In time of recreatión
Nature hath in creatión
The Stomack for a comun coke [1]
Ordeinéd so, as saith the boke:
The Stomack coke is for the hall,
And boileth meté for hem all
To make hem mighty for to serve
The Herté, that he shall nought sterve.
For as a King in his empire
Above all other is lord and sire,
So is the Herté principall,
To whom Resón in speciáll
Is yove [2] as for the governaunce.

And thus Natúre his [3] purveaúnce [4]
Hath made for man to liven here.
But God which hath the Soulé dere
Hath forméd it in other wise
That can no man pleinlý devise.
But as the clerkés us enforme
That lich [5] to God it hath a forme,
Through which figúre and which likenésse
The Soul hath many an high noblesse
Appropred to his owné kinde.

---

[1] *coke*, cook.
[2] *yove*, given.
[3] *his*, this.
[4] *purveaúnce*, provision.
[5] *lich*, like.

But oft her wittés ben made blinde
Al onelich of this ilké[1] pointe,
That her abiding is conjoint
Forth with the body for to dwelle.
That one desireth toward helle,
That other upward to the heven;
So shall they never stand in even
But if the Flessh be overcome
And that the Soul have holynome[2]
The governaunce, and that is selde
While that the Flessh him may bewelde.[3]
All erthely thing which God began,
Was only made to servé man,
But He the Soul all onely made
Him selven for to serve and glade.
All other bestés that men finde,
They serve unto her owné kinde,
But to Resón the Soulé serveth,
Whereof the man his thank deserveth,
And get[4] him with his workés good
The perduráble livés food."

    CONFESSIO AMANTIS, Book vii., pp.
    346, 7. Henry Morley's Edition,
    in the Carisbrooke Library, 8vo,
    Routledge, London, 1889.

---

[1] *ilké*, special.   [3] *bewelde*, master.
[2] *holynome*, wholly taken.   [4] *get*, gets.

90. This extract, which purports to be a part of the instruction which Aristotle gave to his pupil, Alexander the Great, shows that in former ages psychology had an embodied residence in the human mind which it no longer holds. How it originated is not apparent, but it seems to be a survival from the primeval Churches and their Revelations. It cannot have come from anatomy, because it is the living mind that is imported. Yet anatomy of some kind, perhaps of animals in sacrifices, of course opened the knowledge that heart, lungs, liver, spleen are contents in the body. Gower does not mention the Brain, but instead of it the Soul. He lived from about 1327 to 1408. For the above Edition, cheap and elegant, the learned Professor deserves the thanks of all who value English Literature.

Weighing and Pondering.

91. With every inspiration or breath the lungs raise the body in all dimensions, and were it not for the attraction of the earth and the pressure of the atmosphere, they would lift it into their own element, the air. As it is, they weigh it continually in their scales. Also they weigh every organ individually, and make over to the general consciousness of the body its willingness or freedom of expansion. If it is obstructed, or solidified, and cannot dilate to obey the traction of the breath, the fact is registered in defective respiration. All muscular movement and exertion is possible in strict proportion to what the lungs can effect in expansion. Disease and pain proclaim these conditions. A pleurisy stops the breath, and brings its play to a standstill. An inflammation of the

liver, the peritoneum or the kidneys does the same. These are loads that the lungs cannot lift or endure. They cannot be breathed up and down into the general atmosphere of movement.

92. This weighing of the frame by the lungs is also a constant tendency to elevation. The lungs are the mountains of the body. With each breath, the blood that the heart immits into the lungs is raised up into the air, and suspended in it for inspiration. Imagine the clouds of winged blood, the momentaneous crimson flocks and flights on the helpful hill-tops. But not all the birds of the blood can bear the altitude; many fall back wingless into the outer world. The ascent is too steep for some, and they fail to breathe, and sink into the tissues. For they are all weighed in the airy balances, and the flesh of their value is ascertained. The same thing is repeated by the universal lung-

elevators in every organ : the blood is drawn pulmonically by the breathing of the organ into its capillary depths and heights, and there performs its high services to the organ; and also is examined, judged and purified according to the deeds done and doable in that body; whether they be well done, or whether they be ill. Nothing can be weighed without being in some sense raised somewhere. So here we have clearly a universal function of elevation, namely, towards the perfection of the service of the bodily commonweal. Moreover, all summoning of the blood to its intimate capillary houses is necessarily elevation by reason of the naked eminence of the sanctum, and there its quality is discerned individually, and its value is coined and stamped. If the good blood be golden, the heart gives the gold, but the lungs give the currency all its form, and are its mint.

93. When the body is perfectly healthy, and as Swedenborg says, "constituted in motion," its weight is not felt, but we are consciously one, and are insensible of our weight. The heart, our "bosom's lord, sits lightly on his throne." The bronchial arteries raise the lungs to a higher summit, and flesh and body follow them upwards with airy accordant steps. All function or infinitesimal motion then goes on automatically and as it were spontaneously under the grand general motion of the lungs. And then also irrespective of order and function, there is no upward or downward in the frame; we are well in harmony; and the erect body feels in itself no antipodes.

94. Furthermore, although we always breathe *in* the chest, the originant motive of breath may come from any part of the body. It does come potently from any part in distress; the organ or member beseeches the

lungs to breathe down the anguish, and may summon them to utter from the voice groans or cries. We may breathe mainly from and to a wound or a broken leg. So also we may breathe principally and continually in waking hours from the appetites of any organ. We may breathe from and to the things of the belly, the bosom, or the head. We may breathe predominantly from the muscles, and build an athletic life. In these cases the lungs are the ministers of all established and establishing bodily motives. They receive and give wheels and wings, always by corresponding movements, to the whole character of the body. They weigh the body according to its own estimate and adopted centre, and confirm its wills and ways.

95. Thus we have in the lungs a number of correspondential facts showing transparently that the lungs are in the body what the understanding is in the man. The lungs

weigh and adjust according to the integrity and unity of the frame. The understanding ponders according to the sanity or conscience of the mind. The understanding by its constant spirit and affection makes the large mind out of the limited senses, and considers reactively the whole state of the organic will which has the first mind with it. As life goes on it has its own ponderings, for the character is its burden, and often there is "perilous stuff that weighs upon the heart." From mental pain and reproof it is dispirited, and does not know how to think. The duties and delights of life are all sensed by it, and it apportions the worth of performance and the rights of happy days. Especially it ponders the memories, which contain the health or disease of the mind. It knows of broken hearts, and refuses to live for them. The parallel with the lungs is too obvious in these particulars to require to be drawn into further illustration. The trials, burdens and blessings of the mind are here equated in and expressed by the weights and sanities of the body.

96. The tendency to elevation in both cases is equally plain. The equation goes on. The lungs elevate the blood and the body to a high level by their inspirations: as Banning well says, their "plenary inspirations." The understanding elevates the affections towards wisdom by its perceptions and thoughts. On the summit, where the sensibility of conscience is greatest, and the air of its truth purest, it purges the affections by acute self-examination, and casts out their breathings when low or impure. It puts truth in the first place, and stamps it on the will, inviting obedience, to convert the spirit into a reality, and to make good the truth.

97. While this goes on the understanding is itself elevated, and the conscience becomes

clearer, just as perfecting health lightens the body for the lungs, and abolishes clamps and obstructions in the organs.

98. But the motives of life are all; in this case the purposes for which we use our faculties, or the use which we make of our understandings. The question is, what spirit we are of? From this spirit we breathe, live and act; and our understanding is ultimately enlarged by it, or limited to it. In its scales we weigh sensuality against self-denial, and good against evil.

99. It is therefore demonstrated that the lungs in their corporeal functions correspond actually and practically in the body to the understanding in the mind: that the operations of either in their relative spheres tally piece for piece; that the universal secret operations of the intellect can be seen in a mirror from the lungs rightly understood in their uses; and on the other hand, that the

physiological doctrine of universal respiration is elevated by the correspondence into a psychology which actually reveals the connexion between the body and the soul.

100. Shakespear diagnoses the corrective lung-understanding offices with his usual telling arrow, striking the apple on the head. Does he not say, for instance,—

> "And thus the native hue of resolution
> Is sicklied o'er with the pale cast of thought"?

Here the hot heart, sanguine in its first desires, passes its courage through the gates of consideration, and parts with the bloom of its folly in the cool airing of the business-offices of the lungs.

101. We may now turn the psychology upon the physiology.—The understanding in its whole meaning is the eye of the mind, and regarded as the impartial intellect, its separation from the will allows its elevation to any part of the mind whence instruction

can be given. As a mirror it receives the heart or will upon its representative plane, and enables it to contemplate itself in the glass of reflections. When the reflection is placed high, as in the conscience where the Word of God informs, the heart sees itself as from a distance as we see ourselves in a lake or a mirror, and it becomes an external object of thought; a second person; an outer man in the gaze of an inner. So again with Shakespear, the "mirror is held up to nature." And the mirror may be slanted in all directions for fresh points of conscious regard. But if the separation is inconsiderable, or the man does not stand intelligently away from his memorial self, no whole view of the state is taken; and if the intellect is kept close to or under or in the will, it is merely an eye set back in the dark brows of purpose to carry it into effect.

102. Observe, however, that the mind's

reflection not only takes the print of the state reversed, as its office is, and reports the reversal, but also seizes the heart of it in the truths born of reflection, and according to the willing reception of those truths reverses the nature and makes it new. So the understanding, at first the mere luminosity of the will, and the walking shadow of it, becomes at length one with its substance, a new form to it, and in the man grows up into a real human understanding.

103. All this is represented in respiration as distinct from pulsation; in the carrying of the blood up into the lung-vesicles to examine and purify it through the thoughts and perceptions of the air. For the respirations correspond to organic as well as voluntary thoughts, and all thoughts both look and see. In this way the lungs represent an introspection of the body in all its parts, as the understanding is the eye and light of all

things of the mind. So the body is full of its own consciousness, as the mind of its own; and the mental and bodily lights correspond. The union must be close indeed between breathing and seeing; for seeing from thought is conscious breathing: the way of life with its moving events and objects on both sides, in being seen, calls forth exact breaths which are the planted feet of the spirit.

III.

THE HEART.

Centrals for the Centres.

104. AN eminent American Physiologist, Dr. H. K. Jones of Jacksonville, objects to the position that the pulmonary blood in its arterial flower nourishes the lungs, on the ground that this blood is not freighted with its nervous life until it has passed through the heart. There is something in this remark. Yet as the bronchial arteries nourish the lungs, and give them individuality, that is, respiration; and as their fleshly structure is the greatest expanse in the body though the lightest; it is feasible that the pulmonary vessels also contribute to the so-called

nutrition of those organs. Do not the jugular veins carry down from the brain more than the effete blood? The subclavian vein by the thoracic duct carries more than venous blood: it carries the chyle and the lymph, the many times refined essence of the material food, from the kitchens of the stomach and abdomen. Has not the brain its supernal chemistry, by which it excludes, hatches, a new spirit for this new body to convert it into blood, so that the life may be in the blood? Unless by supplying this missing link, this neck of nutrition, from the ever-chymic brain, there is no account of the transmutation of material chyle and lymph into living blood. In this case the right side of the heart, which receives the jugular and subclavian veins, is the rich laboratory of the blood, even more so than the left side of the heart, which is more concerned in distributing the arterialized blood than in making its substance. The aëration of the blood may be overrated as a function if it leads us to neglect the spirit and life that go to its formation. Postponing these considerations,—if the lungs and the heart are in marriage, is it not likely that the great income of the blood should in part be settled upon the lungs as it accrues from the entire estate, and on the day of marriage, that is with the first breath? The contrary seems unprovable, and improbable.

105. This leads us to speak of the centrals of the human system as differently conditioned from the outlying parts; as not mensurable by their rules as we hold them; but furnishing exceptions which, if they are of novel import, yet may enlarge our vision for the schemes of nature. We note this about the brain, the apex and centre of all, where the venous sinuses or lakes have no adequate parallel in the body. In the lungs, where

veins ramify into minuter and minuter blood-vessels like arteries elsewhere, and then, ending in capillaries, their arteries increase from leasts to larger and largest vessels as veins increase elsewhere. In the liver, where veins subdivide like arteries, and again increase, and form a trunk like veins; and carry venous blood all through the two vascular trees planted into twigs, and fruiting into roots. An author has named the liver the pig of the body. But after it has rejected the bile and the gall, its portal blood may be clean enough for the life of the animal, and nourish its flesh, while the hepatic arteries carry on the like function with the bronchial arteries for the lungs, and,—besides enriching the portal blood at the extremities, and empowering the work of secretion,—inaugurating the liver into the respiratory movements of expansion and contraction which it undergoes thirty times a minute, and by which as motory force all its functions are performed.

106. Central events, therefore, do occur in the great seats of power, the organic centres; royalties, autocracies, supersession of rules of thumb and rules of foot. The carotid arteries, to convey a meaning, are no longer flush with their blood like the common aorta, but they climb up by bony spires to the antechambers of the brain, that the blood of the heart may bow before it, and put off the uniform of its familiarity elsewhere. The heart comes under the same aspect of central and supernal governance. For it is the centre of the body. As the Will is neither voluntary nor involuntary, but is only Will, and volitions are its proceedings, so the heart is only blood determined into reciprocal muscles, and it is its free self, and the pivot of itself, and it is its own blood-life.

The Coronary Vessels of the Heart.

107. In what follows on this subject, I make a few slight and cursory remarks, to solicit the reader's best attention to Swedenborg's Chapter on the Coronary Vessels, an abridgment of which will be found in the following pages (n. 142–157).

108. The heart is said by the Clerks of Anatomy to be "nourished" by the coronary arteries, their blood being afterwards returned to the venous side by the coronary veins. Is not this a heartless dole for such an organ? Is it likely that the winepress and vat of the blood imbibes none of its own living wine at first-hand: that it treads the winepress on such conditions? That this muscle of muscles depends for its blood-life and blood-spirit upon small weak vessels given off outside it by its aorta? It is a hard muscle to get into:

perhaps the toughest of all. It is filled constantly with the finest, purest, richest blood for itself and then for all the organs of the head and the body; and can it live by a backway out of the street of one small artery not its own? Does the emperor in his palace with an imperial cellar send to a tavern for his ambrosia? Moreover, the heart is richer in spirit than the purified blood of the lungs; and this spirit is the property of its own flesh. The heart, too, embraces its blood with all its heart with continual pressures, emblems of its own titanic strength. And its chambers, and particularly the right ventricle, are furnished with tridents triple-pointed, with muscular beds and strong contractile caverns, set with great bands and arrests, by which the grips and claspings of the heart are effective. In a word, everything is prepared for the heart to drink mightily of the wine of its own vineyard. There are minute orifices or foramina in some parts studding the ruddy caverns, and which open as mouths polished with drinking, and always ready to drink. L. Langer, a modern Anatomist, says these occur in all the cavities of the heart. The blood also is human, and longs to feed its own form, as the heart hungers and thirsts for it. The lungs in whose breathing arms the heart beats and reclines, assist all these functions by traction. Now Swedenborg's theory is, that both the coronary arteries and the coronary veins, although carrying relatively arterial and venous blood, are veins to the heart, and that they empty into the aorta on one side, and into the right auricle or vena cava on the other: having their terminations in these places respectively.

109. In this view, the heart in regard to its own vessels is a double cone. The proper vessels which animate, or as the saying is,

nourish the organ, are arterial, and Swedenborg denominates them "diminutive aortas and pulmonary arteries." They arise from the little foramina on the inside surface of the heart, they run into the muscular substance of it; they then form channels which become the origins and tubes of the coronary vessels, and these go out to the surface of the heart, and end as mentioned just above. In the systole or contraction of the heart, its proper and intimate vessels send their blood into the coronary vessels of both kinds, which are their veins.

110. The current view is that the so-called coronary arteries come from the aorta, supply and nourish the cardiac muscle, and then ending in the coronary veins, seek the right side of the heart, the auricle near the vena cava. A contrariety of current is implied in these two views, excepting for the coronary veins, which are veins on both showings.

111. In the former case divined, and defended on anatomical grounds, by Swedenborg, the heart wrings into itself what it requires of its own life-blood. There is also a perpetual balance going on in this active and anxious organ. Nature makes equilibrium here for the security of the throne of the body, and makes it by the heart taking in what it wants, rejecting as it pleases, relieving itself of pressures, and so by many compensating channels commanding equality and evenness in the ever-varying populations, interests, and insurgencies of its own blood.

112. In this way, every pulsation, the whole action of the heart, squeezes out the blood of the muscle which the heart is, from the inside to the surface, and thus perpetually fills the coronary vessels. This is an adequate provision for the heart. So are not the Parthian squirts which the retreating aorta, according to the learned, shoots to it. In

this wise it is bathed in its own inward courage, and does not live upon the sweat of its wrestling. The conflict of the two views recalls the war between the old and the new astronomy. It is a battle of centres. Is the solar heart the centre of its own life, or is the coronary planet the centre of it?

113. The gross objection to Swedenborg's theory,—to the Copernican view,—is, that the coronary arteries here classed as veins are made to run into the aorta, and how should veins run into an artery? Veins carry venous blood, and are not in rank with an artery. In the circumferences, Yes; in the centre artery and vein are both extant in a higher name and power. The blood that has just been life to the heart-muscle has no soil upon it but only the certificate of self-devotion and service. It has given up some little life, but it is venous to the royal heart only: for its very deed eminently arterial to the

rest of the body. Often it goes protected by the valves and veiled as a bride into the aorta. It has indeed parted with the prerogative of continuing to be cardiac blood. Yet it is an exceptional blood-angel to the body.

114. Swedenborg's position flows from the following statement carefully considered. "All the vessels of the heart depend entirely on the action of the heart, in the stream of whose motion both they, and the muscular fibres and fleshy ducts, as well as the lacunæ of the ventricles and auricles, are set and disposed" (*E. A. K.* n. 398). Is there in mechanics any thing of fluid, whether water, steam, or oil, which does not flow according to the forwarding and compulsion of the machine? Is there any parallel to the coronary arteries feeding the heart backwards against its structural motions?

115. The subject is too great and intricate

to be further considered here. We have but touched it with common remarks, and insinuated into it, to illustrate it to the reader, a few patent analogies making for Swedenborg's theory. Yet one more thing commends it, namely, its psychological import. For the first time it justifies, mechanically, physiologically, philosophically, the attribution of the will and feelings by endless changes of state and representation to the inner tides of the heart, seconding the character and dispositions of the man in all his moments. The heart becomes a mirror, an expressive face, in the light of this new anatomy.

116. Recognising the need of centrals for centres, and relegating circumferences to their places, we note, in defect of this righteous order, the fate of the Sun himself in the hands of modern astronomers. He, the heart of dead nature, is regarded as the pan

and sweeping of her ancient atomry. His "pure fire," "burning without consuming," and burning with consuming, is cudgelled into him and out of him by meteoric impacts. Cosmos throws stones at him as in a pillory, and so, not nourishes and cherishes him, but provokes him, into warmth. It is like the fate of the heart with the coronary arteries. The mind of science in an age is one, and carries the drift of the time. Materialism is consistent in making exiguous second-hand matter into the centre of all things both in the human form and personality and in the solar universe.

Representation.

117. The face is a living symbol of the interiors, and shows, or can show, what is going on within. Being front on the head, it indicates the presence and pressure of

showing or representation everywhere in the frame. To a mind sufficiently gifted, every organ would have a face; that is to say, its structure and function through its explored visible appearance, and at first always in its place and wholeness, would symbolize, express, and represent brain-qualities and human inhabitation; we should see mind clearly there, and body obscurely. In other words, the end or purposes of the organ in the man would be revealed. The Eye in the elevation of such science would be externally "the light of the body."

118. We have frequently spoken above of this rule of representation, for it is the condition of mental and spiritual correspondence. Yet, though the heart represents the will, and the lungs the understanding, these are but attributes, though real attributes, in the organs. They are telegraphically presented in the heart and the lungs, and not there in

real presence. The heart and the lungs are their theatre. Their life and truth is in the organic mind, and from that mind in the brain and the nerves. Yet were not sentences, scenes, acts, dramas carried on in the blood by the heart, and in the heart by the blood,—things vitally like the mental things transacting in the brain,—in short, representing them by and in their play,—the heart would have no life in the body, or only a diffused existence. Below, realities are carried on by representations, the great Word by parables. This law prevails. The mercy of it to the human mind is, that the higher and highest things can be seen somewise by the lowest faculties, and the mysteries of nature can be opened for them as *they* are opened by deeper lessons and explanations of representation.

119. The brain is the supreme representative and correspondent of the mind; for practical purposes it is the natural mind.

Its cortical substances are little hearts, *corcula*, which enshrine and prepare to incarnate the love and wisdom of the man, such as he is willing to receive them. These are the beginnings of all determinations, the powers of all representations, and their pulses and breaths alone carry with them the nutrition of life. This life they circulate through the medullary fibres, and then through the nerves, to the universal frame. They both animate, circulate, and feed, and are the fountains of the actions of both the heart, the lungs, and the viscera. The brains are of such elevation that they seem not only to represent but to be the will and the understanding. So close is the correspondence here between body and mind. According to the materialists, they secrete thought and mind. This, though upside-down, divines their eminent glandular function; for they do secrete the spirituous fluid which animates the body; and in another set of organs they do secrete a lower fluid which is the blood of blood, and which is the life of the red blood of the system. But the glandular analogy cannot be carried upwards. They do not secrete thought as the liver secretes bile; but the descending soul makes the mind-form, and through the form the brain, and through the brain the thought.

120. At the risk of some repetition, we may illustrate the representative character of the lungs by recurring to the bronchial arteries. They represent freedom from the nature of the heart, although the lungs are its married partner. *Representatively,*—in the corporeal parables,—they are little streams of affection coming off not from the heart, but from the descending aorta, or from small and remote vessels; and thus not immediately connected with the heart, — *representatively*, the love, the Ego, — but with another affection, namely, an affection for the

whole body or commonwealth, which is in the movement of the lungs,—the wisdom,— in the "enlightened" self-interest of *that*. To this affection for freedom the lungs attend and listen *representatively;* and herein they correspond to and are at one with the understanding psychologically. Once this heed and direction given, namely, with the first breath, the way for managing the heart is begun: the heart, as the saying is, is broken in his, or her, wedding-shoes. So this one attention to a new order of influx gives the possibility of respiration and thought as holding with yet not obeying the heart's pulsation. Correspondently, a little further on, the first received instruction of the selfish little infant in the mother's lap,—the first " Naughty " from her loving lips,—begins the commandments and the Gospels. An infantine awe, a whimper, some small seeming break of love, and a little remove from it,—something of bronchial artery,—begins a second life, and baby begins to know what "enlightened self-interest" is; for milk may at length depend upon giving in. But education, instruction go on, and work to modify some of the self, and in long time are adopted and acted upon, are perceived and are seen to be indisputable and indispensable. Yet the self, the original odd nature, the *proprium*, remains for ever at the bottom, only more and more plantar, or nearer and nearer to the footsole; just as the bronchial artery remains for ever a dependent of the heart, but though its apparent minister, its real lord. It is not, however, an immediate dependent, but still always partakes of the quality of the blood,—the life-truth,—be it noble in its capacity of representation, or ignoble. Such appears to be the small hinge and intercalated pulley,—the to us imperceptible "clinamen" or inclination,—by which the lungs swing free and take the air; opening the door of a greater world than the heart of itself knows; and afterwards introducing the heart with all its blood to that immense horizon where view is added to existence. Out of this comes the state of Freewill, and the consciousness of it, though this faculty as the essence of man is indeed a perpetual divine gift only to the will in conjunction with the understanding, and exists merely by representation in the breast itself and the organs and members below it. For, as we said before, the body is the theatre and the play, the mind is the personal will and understanding. Yet the body,—a body for ever,—is fundamental; for apart from its powers of show, manifestation, ritual, there would be no ultimatum, no reagent platform for these faculties, no senses and no actions.

121. Thus there are two selves represented in the heart; two Voices: the thoughts

K

of the heart alone, and the thoughts of the heart with the lungs. Both are indeed with the lungs because all thought involves respiration. But in the first case the heart rules, in the second case the lungs rule. Sometimes a man thinks worse for disregarding his first thoughts. Sometimes " he thinks better of it " when a first wilful thought has carried him away. Second thoughts are sometimes best. Consideration like an angel comes and chases the offending Adam out of us. The better self in a man may lie in his heart by good parentage, or come into it from without. The poorer and leaner self may be there in dry intellect and self-consciousness. Whichever way it is, both selves are *representatively* of the lungs with the heart, but differenced according to the height breathed from, and ultimately according to the predominance of either organ.

122. The bronchial arteries as ministers of

respiration may stand in the mind-series as the rational arteries, and the pulmonary arteries and veins without the bronchial, as the passional arteries. The whole circulation of the blood, could it be regarded separately, would be a passional circulation. The lungs signify all law, and the heart signifies all liberties. Union must exist between these two, or neither can endure under those good names.

The Mental Heart.

123. If the heart represents the man's love or will, it may be possible through a true intuition of the organ, without going very deep, to see some common points of the symbolism from one's own inward experience and conscience, or from knowledge of the world, applied to the nature and ways of the heart. It feels with the feelings and

emotions—sympathizes—but this does not point ocularly to structure, but to general location. But can we read love, will, self in the shape and make and work of the heart in images or mirrored forms :—in the business it is always carrying on ?

124. Let us look first at the proper circulation of the heart. It takes care of itself at once from the finest of the blood and spirit, and from the firstling flocks of the food driven in from the plain under the jugular mountains. These fluids,—for all is liquidate here,—come to it with varying velocities. The centres of the currents are swiftest and purest, the sides are more and more sluggish. The centres are caught according to quality by its fleshly columns, and forced into its caverns, cellars, palaces, temples, whither it wills. For these the heart is self-contained. The heart and they take no heed to the great stream for the body rolling on through the

arterial tunnels to the lungs, and from the lungs again to and through the heart. This stream is like a wall which serves as a cabinet for their privy councils and intimate decisions. Streams in fluids, as marine science knows, are barriers which currents bent to other destinations do not infringe, but pass under them, over them, and beside them. Ocean currents make the sea organic, fibrous, muscular, and tendinous, from the equator to the poles, and from the poles to the equator, tubulating the great deep as into arteries and veins. So here this Red Sea is as a shore on both sides for the perpetual passage of the purest bloods, the chosen people of the heart.

125. The heart is indeed a great example of outpouring and giving away through the general arterial system. It gives all its blood away at last. But in each stroke of its life, it serves itself first, and with the best

life, and gives the second best to its neighbours, the organs. It needs the best for its central work, and is justified in taking and appropriating it. We may call it the very organ of property or proprium. It is all muscular, oneness of effort, and by the compaction and look of it, listens to nothing. It is representatively a will that stands above consciousness, and reveals itself by general and not particular states, or only by striking upon the ribs when it is offended. Being of the form and shape of the love or will, we do not pour will into it, but it pours will into us, automatically and unconsciously. Yet, as the will is the whole man and the sole man, so this curious and most curious self, hidden in its reality from all but God, hidden from the man and not hidden, is perceived by his fellows through his person, life and actions; and himself is interpreted by it. The world does not see auricles and ventricles, but it always sees and says ultimately that the man's heart is himself. So again the perceptions of language, perpetually inductive, as a heart-voice from other men are seers of the heart.

126. It appears now in this physiology of the circulation that the heart is the most grasping of all the firms in the commerce and traffic of the body, and of itself knows itself only. In the current physiology, it sends all away from itself into the lungs and then into the body, and is a mere muscle showing no representation of that self-love which is the permanent low centre of the life of every man.

127. As we saw above, the mental heart, the selfhood, is a natural claimant for miracle ; without this it never alters itself. The Lungs in their elevations are the means for the miracle. The prayer is, "Lord, give me a new heart and renew within me a right

understanding." The heart is to be born again. This cannot be represented and seen physiologically; for to the bodily eye the heart of the righteous man shows no characters to distinguish it from other hearts.

128. But to resume our matters. After the heart has made thorough use of the choicest blood, its own proper heart-blood, it sends it on, by forcible embrace ejects it forth, into the coronary vessels, partly 1, for recirculation in itself for all kinds of mediations, accommodations, introductions, amities, heart-purposes; and partly 2, to go into the general circulation through the aorta, and into the pulmonary circulation through the pulmonary arteries. The first kind helps to make all the blood suitable to its own nature, character and constitution; the heart being the compounder and differentiator of the blood, drilling it out of chaos into order. The whole volume becomes unitary and

single-hearted hereby. For this subject, see Swedenborg's light on the coronary vessels. What the heart sends out into these on the arterial side, bears the messages of the heart, affecting and effecting its nature and bequests. The coronary arteries, to attend to them alone, represent affections of the will or love like all the arteries, but being immediate emissaries in the heart and from it, they represent the flesh of its selfhood more nearly than the aorta, the blood of which is comparatively its common and not its intimate minion. The general circulation through the heart knows it only as a master of many strokes driving it elsewhere : the blood imbedded in the heart, from the columns, caverns and foramina to the muscular fibres, knows the heart as its intimate flesh, and preserves the memory of it. With this experience and recollection behind it, the coronary blood carried into the aorta is charged with the heart's ruling peculiarities and traditions in a higher degree than any other blood that is pulsed from it. Blood that goes from an end to an end, is not promiscuous like water in a bottle, for it is full of will and cannot be shaken: it is of its nature *a priori ad posteriora*, purpose-carrying and purpose-seeking; and can be an undisturbed traveller in an ocean, full of the past and the future, personal as a Captain on a chartered voyage.

129. Thus we may conjecture that the heart in its coronaries, has attached ministers proceeding to the body, just as the love and will in the mind have reporting affections, pleasures and friendships far away from the centre, which feed, defend and extend the *amour propre*. By virtue of these the ruling love is warned of whatever may infringe its habitudes. Self can take care of its throne: these look to its furnitures and footstools.

130. For we must look upon the blood as itself the body, in the sense of being organic by priority to the body, though not without the body. In this view it has sympathies from part to part, fluid members and organs in social union, and all these have one soul. So *per contra* the coronary vessels debouching into the aorta are model instances for the purification of the blood in the liver, pancreas, spleen and kidneys, as pointing a high example of elimination,—of the secretion or self-denial of the blood itself; and this blood especially in the kidneys may be the judicial agent in the renal stream for sentencing the fluid of the kidneys to extrusion. The Scriptural word, "Searching the heart and the reins," leads to this conclusion. The function here is of the heart united to the lungs, of the love purified through the informations of the understanding.

131. The heart rules towards the pulmonic

circulation, and until the blood reaches the lungs. So the love rules in the mind until the understanding lays hold of it by thoughts, puts them into it, imbues it with them, and takes it to pieces in their analytic freedom.

132. The heart reigns in order in the body throughout, married by nature to the lungs. So the love or will reigns in the general mind, but the understanding by senses environs and informs it. The minutest arterial vessels, and the capillary extenses, are under the law of the lungs, but from impurity of the blood are subject to blocks and obstructions which it is difficult to breathe down into equilibrium, and so to continue in the lubricated wheel of circulation. So in the mind, the small oddities of self-love stand out, and tend to bring back the great things, and to reinstate them, using the weak places of the understanding to defeat its common faculty. These are beginnings of disease

in the body and the mind. For diseases of the body traverse the respiratory power, forbid plenary inspiration and expiration, and subtract muscular strength. They either thicken or emaciate the organs and tissues, or make the muscular case with the ribs weak, and thus the organs fall into displacement and breathlessness. The mind also totters correspondently, and thought lacks animal spirit to stand upon.

133. From the principle that the arteries proceeding from the heart represent affections diverse according to the organs to which they go, it becomes possible by the correspondence of the organs to give names respectively to the affections. Thus, as the lungs correspond to the intellect, the organ of truth, the pulmonary arteries correspond to the affection of truth, and to its perception and regard. But again the bronchial arteries in the lungs correspond not only to an affec-

tion for truth, but to an affection for the freedom of self from self by the truths received in the understanding. This affection is necessary before the love and intellect of truth can come into play. It therefore divides or splits the selfhood into two parts: one which carries the old inherited nature, what is generally called "human nature;" while the other carries a new considerate *renaissance*. The selfhood thus becomes twofold, external and internal, and the external or bronchial may rule, and then it becomes the internal. It is, as we have often said, necessarily rooted in self, for all love, affection and action must have that ressort or they would not be our own; but now the selfhood has two ends, the lower springing from self-love, the higher coming up into the proper mind, the intellectual lungs, fastened there, if we may so speak, to heavenly love, which is love to the Lord and love of the neighbour. At the upper end wisdom has met our ruling loves, and been received; a wisdom infinite in details; another name for conscience, which oversees every drop in the cup of desire, and purges away its poisons. Our last perception about the bronchial arteries may therefore be, that they represent the progress of that pilgrim, the mind advancing in regeneration, from the ground of "enlightened self-interest," social, politic, economic, moral, towards "the hills from whence cometh our help;" the pure unselfish faiths and affections of love and charity and usefulness which are the spiritual man.

134. As the aortic arch is a firmament over the blood of the body, it corresponds to a general affection for the natural life itself; and in the spiritual world for the spiritual life: for the immortality of the organs and members. The carotids going to the brain correspond to affection for wisdom. The

arteries supplying the stomach and bowels, which digest the food, and raise it towards the blood, correspond to the love of continuing life itself as daily bread for use and memory. The hepatic arteries minister vitality to the turbid current of the abdominal veins full of the grossness of the belly; they may correspond to affections of courage going to and through the conscious unclean mind, to enable it to bear its reproaches and cast their bitterness out: their mental gall and wormwood. The splenic arteries give rest and play to their blood, give it the correction of exercise: the affections corresponding are for respite from fear and care; for a quiet mile on deck in the sea of self-consciousness. For the digestive system can be a factory of cares: of doubts whether offered good things can ever be appropriated. The arteries going to the kidneys, themselves in a great strength and centre of motions, and which summon and eliminate the impure serum from the blood as the coronaries discharge the heart-used blood itself, correspond to affections and passions which deal with old thoughts, and set them aside ; which break up consociations of false ideas ; thereby clearing the mind's constitution. Enough, however. We speak of the affections on the good side. They may however be designated on the reverse side according to the life led in the body.

135. In all cases add the correspondence of the lungs or understanding to that of the heart or the love. Again we illustrate this by the kidneys. The blood is brought to them in the renal arteries to have the waste serum separated from it. They are searchers and strainers. But they have to be put in action as a sieve in the hand. Of themselves they have no action but a form

prepared for action. The arteries give them no kidney-life or function. From what hand does their work come? From the brain and the lungs. They are drawn into an importance of dimension not their own, but communicated from above. Respiration, tendency to vacuum, comes to them, not in breathing air, but in breathing blood; in breathing in arterial blood, in breathing out venous blood, and in eminging stale serum. It comes by the universal pulmonary motion sensible everywhere. This pulls the kidneys out, and the exactly attracted blood rushes in to fill the void: then, the inspiration ended, they subside, and the weight of the body, the general pressure, contracts them, and elimination is performed. Every organ is a muscle for contraction, and gravitation itself is *quasi* muscular and nervous, pulling things together. We persist in dwelling on this neglected respiration because it is so intimate and so unacknowledged. Powerlessness to act without the brain and the lungs applies to every organ. Without these two the heart could not exist: without the lungs it is a sleep of the palpitating flesh. After birth it is not awake of itself to its functions. So the affections of the love without the wisdom, of the will without the understanding and its five senses, of the passions without conscious motives or thought-forms and imaginations, have no moment of life for the general mind.

The Circulation of the Heart and the Coronaries: their Correspondences.

136. The heart is the artery of arteries and the vein of veins. First, its great cavities are this. Then its proper arteries, as we have seen, are the ducts which run direct

from its blood into its many chambered muscles. These Swedenborg calls its little aortas. They correspond to the close loves of the heart. The love of all life, bodily, natural, spiritual, celestial, according to the man's character, is representatively habitant in them; grossly speaking, the legitimate love of one's self for one's stable place and use in the Creation. The divine influx into them is for this end. At the venous side the corresponding vessels, which Swedenborg designates little pulmonary arteries, minister the same life, but to be disciplined by repeated ordinations on the way to the air-world or thought-world of the lungs. The heart does not live from affections, but from loves as their fountain-heads: it is itself a substance-man of nothing but loves or heart-natures, and all affections proceed from him.

137. The Coronaries on both sides corre-

spond to the direct insinuation of the universal cardiac life into the general and particular life of the great systemic circulation. But for their streams from the love-heart the universal end would not be represented in the torrents of the body. The coronaries are gulf-streams which warm the poles of the organic planet, bodily and mental. The great aortic stream in this comparison is rather a lung-stream than a heart-stream. It knows the heart as a Lord, not as an intimate Friend. The coronaries know it in the latter relation, and are its dear domestics. In this respect they are affectionally and representatively the first helps in the general system to the universal, general and particular ministrations of the lungs which they are soon to feel and to obey. They convey the absolute heart-world into the absolute lung-world. Thus carrying great ends and their fortunes, in all health their good streams are safe, and cannot be drowned in the other rivers of life.

138. The circulation of the blood as a doctrine is not only twofold, as Harvey left it, but it is fourfold and manifold. First there is the systemic circulation as known to physiologists: the circulation from the left heart through the arteries and the capillaries, and back through the veins to the right heart. This corresponds in its streams to all the genera and species of the affections for the human form. Next there is the pulmonic circulation, from the right heart through the lungs to the left heart. The great atmosphere here intervenes by weight and pressure, and by chymistry: correspondentially the mind, the understanding with its senses, meets the heart, marries it, and forms and reforms it. The third and fourth circulations are cardiac, transacted within the heart itself,

from the openings in the heart to its muscles on both sides. These, moreover, embrace minor subordinate circulations. Such is a general view of the several circulations and their correspondences.

139. Our "modern thought" leaves out the lungs as spirit and the heart as flesh. This it does by confining the lung action to breathing air, thus to the confines of the chest: and by reducing the heart to an organ of circulation external to itself, and omitting its circulation within itself. But while the lungs breathe air in the chest, they breathe body, the whole body, beyond the chest. A singular phrase, breathing body, but veracious, for they breathe and pull the entire frame up to them, and this is the general name of all its functions. Then, again, the heart is relegated and confined to the lung-neglected body, and has no flesh for itself. *Sic vos non vobis mellificatis apes.*

Whereas the heart must live for itself first in order that the body may live at all. We have, however, sufficiently seen that the heart is no mere system of tunnels, but that in itself it is peopled with endless *dramatis personæ* of central natures or affections. Also that the lungs are not mere air-ways or bellows, but have the whole body in their keeping, again to dramatize it for the use of the moving soul.

Relations of the Brain before and after Birth.

140. During embryonic life, the brain which houses the spirit about to be incarnated, namely, the nascent will and understanding, under the Creator designs and constructs the body by the foremanship of the heart, and the strokes and pulses of the heart are its commands and skills. The

inner room of this creation is strict and private, with only space for the workman. No air is admitted to the blood, and no void room environs the fœtus, which swims and grows in a little sea of its own. Muscular movements after a time betoken blind affections of the heart-muscle, ante-natal stretchings. The brain is in the design of this, and as the advance of design is impossible without movement, so the brain is in motion correspondent to the motion of the heart. The brain does what the heart does, but in a higher sphere, as an architect creates what his clerks are to do, but in intellectual lines and limnings, not with stones and bricks. In a word, the animations, or voluntary and intellectual brain-strokes, are now synchronous with the pulsations of the heart. What the animation prescribes, the pulsation carries out. The throb of the arteries is architectonic from the thrills of the light of the brains. The brain correspondentially is thus both heart and lungs to the fœtal body. It not only builds, which is the heart-function, but also inspires the building, which is the lung-function. The heart is subservient and obedient. The forming structural will and understanding are in mere unition at first. The heart is in the stream of the cerebellum and the cerebrum.

141. After birth, the administration of the body throughout life, including growth and development, is the business of existence. For this second course the heart is an independent organ. It is itself, committed to itself, full of the heredity of the person, subject to all the motions and perturbations of the mental faculties, to all wills, desires and emotions. Brain and head are above these things, and can in nowise be in rhythm with them. So it is that in the profound harmony of the womb-life, the heart and the

brain must be at one, or formation would be impossible. In the perturbed life of the infant, the youth and the man, the heart and the brain must be separated, divorced from synchronism, and another marriage must take place. Otherness of life involves otherness of movement, and the new movement and marriage is an animation or rising and falling of the brain synchronous with the respiration of the lungs. Therefore, coming down into a new life, the brains change the way of their steps. We know of the pulsation of the fœtal heart, and may be sure that if the brain before birth coincides as of one mind with the heart, it animates in keeping time with the pulsations : another way of saying that it precedes and urges on the structural formation which is done by the pulsations. We know of the respiration of the lungs, and may therefore be sure that if the brain after birth coincides in thought with them, it animates them by keeping time with the breaths. Yet the spirit and the body are reverse as day and night. For when the body is expanded by the lungs, the brain is contracted, and *vice versâ*. So the brain gives away and sheds its spirit in and with the momenta of the lungs opening to that spirit, and taking it in, which is effected in every part of the body as it also opens to the pulmonic inspirations.

Swedenborg on the Coronary Vessels of the Heart.

142. The first intention was to print Swedenborg's Chapter on the Coronary Vessels as an Appendix, but it seems better to refer Anatomists to his *Economy of the Animal Kingdom*, rather than confront the general reader with a long technical discus-

sion. The alternative is to give some general abstract of the subject as Swedenborg has left it. And first we premise his general Induction of the whole subject, as follows :—

143. "The coronary vessels of both kinds, arterial as well as venous, arise from the heart itself, and not from the beginning of the aorta. For there are little columns and lacunæ in the ventricles and auricles; there are fleshy ducts, and there are motive fibres. The blood flows from the heart into the lacunæ, especially under the columns; from the lacunæ it is expressed into the fleshy ducts; from the fleshy ducts into the fibres; from the fibres into the coronary vessels, both arteries and veins; from the coronary vessels, either through two foramina into the aorta, or through one large foramen into the right auricle, or through several foramina into the same: but the superfluous blood in

the coronary vessels runs back into the lacunæ and ventricles.

144. "All these vessels depend entirely on the action of the heart, in the stream of whose motion both they, and the motive fibres, and fleshy ducts, as well as the lacunæ of the ventricles and auricles, are set and disposed. From which it follows, that all the vessels that occupy the surface of the heart are venous, the arteries corresponding to which are in the substance of the heart.

145. "Such then being the origin of the coronary blood, it follows, that the superficial vessels, commonly called coronary, perform their diastole when the heart performs systole: and in like manner that the superficial vessels of the auricles, perform their diastole when the auricles perform systole, and *vice versâ*. But as many anomalies occur in the auricles, and in the right auricle

particularly, so, in order that the auricular blood may find an outlet in all cases and under every circumstance, a number of orifices are provided, through which this blood can be thrown out, suitably to all diversities of state.

146. "If we compare the origins of the coronary vessels with the outlets of the same, it will be evident that the blood of the right side of the heart is transferred immediately into the aorta, and the blood of the left side of the heart, into the right auricle; much as was the case in fœtal life by means of the foramen ovale and ductus arteriosus; showing that the coronary vessels and their mouths, relatively to the determinations of the quantity of blood running through them, are substituted in place of the foramen ovale and ductus arteriosus; the channel and mode of circulation only being changed. And this, in order that the venous blood, in this place

of concourse, may not injure or destroy the natural state of the kingdom, subject as it is to such frequent mutations.

147. "Hence it is clear, that neither the motion of the heart, nor the circulation of the blood, can subsist for any length of time, unless the peculiar vessels of the heart that discharge the blood into the aorta, and those that discharge it into the right auricle, pursue a perfectly distinct course, and have no communication with each other. Were they conjoined, the same effect would ensue as if the septum between the ventricles were perforated.

148. "Meantime, whoever attentively examines and considers the origin, progression and outlets of these vessels, will see in them, and consequently in the heart, an image and representation of the state of the body and animal mind. In which respect numerous affections may not improperly be attributed

even to the heart, according to the usage of common discourse."

---

149. The anatomists upon whose descriptions this theory is based, and especially Morgagni, record with no exception that the orifices of the coronary arteries in the aorta, one or more of them, in a proportion of subjects, open into that vessel, *behind* the aortic valves, and into these orifices the blood cannot enter when it is propelled by the heart through the aorta. This perplexed Morgagni, who could not say when or how the blood got into these orifices. And upon this Swedenborg remarks that "the law of nature, constant in its causes and effects, forbids our attributing to one heart, or to one of two orifices in the same heart, what is evidently denied to the other." The modern text-books of

anatomy take no notice, so far as I am aware, of this difficulty, but place all the coronary orifices above the valves.

150. Now "Anatomy forbids us to conclude that the blood is conveyed by the aorta through the coronary arteries to the surface of the heart. For when the heart is constringed and the aorta expanded, the blood cannot pass thither through either orifice, because at this time the valve is carried up, and the orifices are closed by its outspread curtain. Nor can the blood pass through either orifice when the heart is expanded and the aorta compressed, because at that time the torrent is driven onward through the arteries, so that none but the superfluous or refluent portion can regurgitate into the orifices; which it must do without impetus, or power of forcing itself in. The heart itself, on the other hand, demands an immense force to throw the blood into it,

which also the arterial coronaries cannot supply, because they are distinguished by no muscular circles, but by little wrinkles and folds of fibres."

151. "Furthermore," he says, "that to grant to these coronary vessels only two inlets, and not always even two, into the aorta, would be to expose the heart itself, which is the purveyor of the blood to the whole body, or rather the muscle of the heart, to dangerous hazard, and to suppose that it holds its life by tenure from its own artery; when yet nothing demands more present abundance and supplies of the blood, which is its own property, than the heart."

152. How the blood is forced, wrung, squeezed into the heart by the heart itself, is shown in Swedenborg's premised Induction given above, every point of which has a long consideration devoted to it in his Chapter.

See especially the function attributed by him to the rugæ or folds of the vessels.

153. He also sees the heart itself as a stupendous theatre of currents or channels of nature, which are different for every heart. It is a great community of bloods and sanguiducts in which all the vessels have a common or general motion, and also a perpetual intercommunication. But the coronary arteries and veins so called, which are the terminals of this system, do not intercommunicate, or had not been shown to do so in his day.

154. In characterizing the channel-work of the proper blood of the heart, not meaning thereby the channels of the pulmonary and general circulations, Swedenborg assigns to intra-cardiac currents the names of refundent, retorquent, anticipant, transferent and retroferent vessels.

155. " If the field of least vessels," says he,

" be that where nature most especially plays, and celebrates her animal sports, and if those vessels depend upon the governance of the brains more immediately than the larger; and if in the field of those vessels changes occur according to all the actions and affections of the brains, and which through the intermediate arteries affect the pulse of the heart; and if the least vessels be considered as placed in one extreme of the sanguineous system, and the heart as placed in the other, maintaining a mutual relation through the larger arteries; and finally, if the heart itself be encompassed with a similar field of least vessels; then it follows that there is no change arising either from the brains or the body throughout the whole sanguineous system which is not represented in the heart. When the *transferent* vessels are multiplied beyond their due number, or open too widely, they represent a weak, timid, unsteady state of mind, and a corresponding state of body. With these transferent vessels the *retorquent* vessels of the left ventricle concur, and the *anticipant*; in fine, all those called arteries. On the other hand, the *retroferent* vessels commonly called veins, if very numerous and open, indicate a firmness and strength of the nervous, arterial and venous systems; and with these the *retorquent* vessels of the right ventricle concur. But the *refundent* vessels both of the right and left auricle, when they are present and multiplied and expanded beyond their due proportion, are significant of frequent changes of the body and animal mind; of irregular pulses, frequent inundations and palpitations. . . . The appearance of the heart depends principally upon these [coronary] vessels. . . . Hence we may see how rude irregular motions and impulses occurring from day to day, deprive the machine of the heart and the entire sanguineous system of their integrity of state, forcing open passages which ought to be closed, contracting those which ought to be opened, and thus inverting the proper order : how by these means a nature is at length superinduced which rushes with blind instinct into the lusts which necessarily result from an altered fabric of the heart. . . . Thus we see a covenant of perpetual amity established between the brains and the rest of the body through the medium of the heart."

156. "Each particular viscus of the animal body has its own science of angiology, or its own particular doctrine of arteries and veins."

157. Swedenborg gives a second version of the heart in the *Animal Kingdom* in his Chapter *On Organic Forms generally*. His works contain hardly any other notice of the subject. " The heart itself," he says, " is an

organ for preparing liquids for the composition of the blood, and at the same time is the beginning of the circulation of the blood; for the blood is not only worked about and digested in the heart, but is also discriminated into parts; in short, the purer essence is driven out into the coronary vessels, and the grosser is sent away into the lungs. The part that is transmitted into the coronary vessels flows out into the beginning of the aorta." And in a note he observes that all the coronary vessels are "the veins of the heart, and are on the same principle as the bronchial vein in the lungs, and the hepatic vein in the liver."

158. These brief extracts must suffice. Swedenborg, a combined anatomical and psychological Seer, has represented the heart as in itself and for itself the principal organ of the body, and the express representative of the lower or animal mind, and united to

the lungs, of the higher or rational mind. This alone gives his remarkable and novel theory a claim to be first studied and understood, and then considered against the facts of anatomy, to see whether it be indeed a truth. And here we may remark that a psychology may be true and valid, and not be represented in any adequate knowledge of anatomical facts. Psychology in the classic ages was in this condition. See John Gower's *résumé* of it. This in no way disproves or discredits the psychology; but shows that it has now to wait for a rational scientific foundation. Not having this, it may pass away for a time, as the attribution of psychical and affectional states to the heart has passed away. The psychology, in order to descend as thought-sight into the organs, depends entirely at first upon a *general* anatomical knowledge of the organs, such as existed well in Swedenborg's day.

The Veins.

159. The veins seem to have less physiological thought spent upon them than the arteries; like woman in history: their record is short: when the arteries are functionally described, it is enough to say that the veins accompany them, bringing back the exhausted blood to the heart. They carry our mortality, and as we die momently, their work is extensive. But the last word cannot easily be said about them.

160. The blood after passing from the arteries through the capillary system enters the veins. It has now lost its arterial redness, and is blue or dusky. Its chief works seem to be done in the capillaries. The affection with which it makes for each special organ is an end of particular service: for the body consists of nothing but ends or final

causes in circular and spiral series of series. The ends are to give life to the organ, which is done by the self-sacrifice of the blood which is the life when it comes into its individual globules in the capillaries. Thus the individual globules, like all individuals, have each the blood soul in them for duty and love of the organ which is their neighbour, otherwise than when they are in rushing masses or crowds in the arterial streams. In the capillaries, through the respiration, they animate the organ; they repair and rebuild it, for they are transmutable into its solids; and they purify themselves by secretions. It is of their nature that they can become in time all the organs through which they circulate. With their forms,—with the brain round them, and the life in them, and the lungs drawing them on,—organic architecture, organic material, and organic purpose and want for the bodily man, are everywhere immanent and urgent. Under these auspices, as kidneys are essential, the kidney blood first builds and continually rebuilds them. It has the eidōlon and then the idea and then the workmanship of them in every drop of it.

161. But when the end or ends are accomplished, the blood is comparatively at leisure and rests. Ends on their way burn with the love that is in them, and are ardent and flaming: in their work they are settled, not hurrying to the place, but at the artisan-bench; making kidneys, or feet, or what you want; they have the common hue of the workshop: may we say, they are unnoticed and colourless; private, and stick to their last. When the end is discharged from them, and the individualities are again massed, they have lost one life and its fulgence, and are under mere pressure in the veins, and not propelled by pulsation. They

have valves to crutch them, and prevent them from falling back. They cannot tumble beyond a safe number of steps, and even then into nets or curtains.

162. Yet the blood in the veins is not less pure than the blood in the arteries. The blood in the renal veins might be reckoned more pure, for it has parted with the urine in the kidneys. The blood in the ascending vena cava has parted not only with this but with the bile and the gall. We may infer that the venous blood after these purgations and sacrifices is old and weak and hungry and appetent, but not otherwise impure. The end in it is changed. It is going to a new kingdom and a younger progeny in which it will live again.

163. The venous blood returning from the brain is especially lustrated and purified, both in the skull, and in the organs of the senses. This is evident from the arterial

secretions of the eyes, the ears, and especially of the nose. The mouth and pharynx also are a vast field of sputa and mucosities from which the veins are freed. So also are the trachea, and the bronchi. In the mouth the appetizing saliva is given away by the arteries. The blood in the superior vena cava especially is emptied of impurities, and weak in itself is led onwards down the beneficent neck to its new career. It is a proof of the comparative purity of the venous blood when it is in the right chambers of the heart, that it is immediately committed to the lungs, and the light winnowing air in them is the last trial it has to undergo, and the final acquittal. Very different from its lower sentences and fines in the liver and the intestinal tube.

164. The venous blood streaming onwards to the right side of the heart, and with all these impurities drained and strained from it,

is gifted with the new chyle mingled with abundant lymph from all the organs, by the channel of the thoracic duct, which at the bottom of the neck enters the junction of the left jugular and subclavian veins. This chyle is the realized nutriment that has been afforded by the food taken, after passing through the examination of the lacteal vessels; and it also contains the returned lymph from those higher veins, the lymphatics, and which has passed through the lymphatic glands. It is the contribution of the body to the blood. If it were the whole account of the constitution of the blood, then brain and nerves, and the complete man would be nothing but body, and the Lucretian dogma, (*De Natura Rerum,* Lib. II. 864–881,) would be partly just, that we are made of what we eat, of flocks and herds and vegetables. But Swedenborg sees that the venous blood from the brain has endowments of a noble order from that summit of life and activity in which all function is eminent and supreme. The brain itself is almost fluid, and in its ventricles and interstices abounds in shallows and lakes. Is it possible that what is thus admitted to, or emitted from, the brain is on the same level of function or structure as the lubricating fluid of the lower serous membranes? Common sense contradicts this, or if not common sense, the first blush and probability of it. Perhaps Chemistry, always working with things dead, or with after-deaths, can detect no difference between brain-lymph and body-lymph, but then Chemistry must have Life before it speaks: and life is in the nature of things beyond its pale unless the chemist himself spiritually alive imports it.

165. Professor Tafel says: "According to Swedenborg, there are three general functions of the brain—(1) the function of sensating

and perceiving, (2) that of determining and acting, and (3) that of conceiving and bringing forth the nervous fluid or animal spirit, and the blood."

166. "For the purpose of preparing the blood," says Swedenborg, "the Soul has established in the cerebrum an illustrious chemical laboratory, which it has arranged into members and organs. and by the ministry of these it distils and elaborates a lymph animated by the animal spirit, whereby it imbues the blood *with its own inmost essence, nature, and life.* This is the object of certain organs of the cerebrum, namely, the corpus callosum, the fornix, the three ventricles, the choroid plexus, the glands or tubercles of the isthmus, the infundibulum, the pituitary gland, the cavernous sinuses, and several other organs. The essential animal spirit must be in the blood in order that it may act the part of a corporeal soul,

or play as soul in the body. That there are two natures in the blood see *The Animal Kingdom*, Part I. The animal spirit cannot be poured into the blood-vessels, because it is most volatile, unless it be conveyed thither coupled with lymph. Therefore this lymph is the very purer blood. So the cerebrum is a great conglomerate gland, to which as a conglobate gland the pituitary gland corresponds." (*The Brain*, Vol. I. n. 90.)

167. "The lymph of the cerebrum is carried down by the jugular vein, and the chyle of the body is conveyed by the thoracic duct, in such a manner that they meet one another, and produce [in the neck, the place of universal *nexus*] that noble offspring, the blood." (*Ibid.* n. 93.)

168. The above conglomerate of organs, the uses of which seem to be unknown to physiologists, Swedenborg regards as a separate system, and speaks of them as THE

Chemical Laboratory of the Brain. The animal spirit is the first product, the white blood is the second, and becomes the vehicle of the first, and both then enter into the new red blood which is the product of the union of the animal spirit through the lymph or purer blood with the chyle. The jugular blood is the bed and bridal, the offering and fortune of the parents for the new lives.

169. Can we get a clearer meaning of Spirituous Fluid and Animal Spirits to conciliate the metaphysicians and physiologists? Whatever carries an end or purpose of life with it, is in the spirit of that life. An apple pip is instinct with the force or spirit of producing in and through the ground and the air, and through heat and light, an apple tree, and apples. That is its evident end. The pip knows nothing about it, yet does it. It has a soul of good use, but not of intelligence. It has no brains.

170. The animal spirit has brains continually pressing through it, and making it alive for the ends of the body. What is called protoplasm, the stuff which can convert itself into so many organic forms, is nothing but a vehicle of intelligent ends, all of which emanate only from brains, either from the conscious cerebrum, or from the unconscious cerebellum. The unconsciousness is no difficulty, for we are unconscious of our own souls, which are ourselves in an intimate degree. Wherever therefore any organic end is carried out, as everywhere in the miracles of the body, there is a living spirit that does the work, an animal spirit; and in the head or the body it must be of the brain or the body; it must have a cerebral or a corporeal vehicle; it must either be a fluid or a solid. Otherwise it could have no contact with our frame. There is therefore, while we live in bodies, nothing in the objec-

tion that spirit and fluid and life and fluid are incommensurable entities. Suffice it that we are incarnate and incorporate, and that we are also alive. And in short, as we have said repeatedly, every part of the organism is human in its ends, and clothed with a humanity exactly adequate to the effect required. If blood-making is the effect, the man-spirit from the brain ensouls, the man-lymph from the brain receives and clothes the soul, and the man-chyle embodies into corporeal blood. So the blood has the life in it, and can work the man-soul's commands or ends wherever it goes.

171. This embodiment is wonderful to contemplate. All the most subtle and dangerous principles in the world are taken and tamed by it. The imponderables are there, electricity and magnetism, fire and light beyond mortal conception: the substances and creatures of every science. All

chemistry and all mechanics. Polarities to new ends. All suspended upon unimaginable threads and films, in ovaries and wombs in which they are reborn, and new named. And incomparable safety is in the incarnation and the castle of it. The self-preservation here bespeaks the soul. Only our own passions or evils can burn us down or break us up. Had we no weakness from these, lions and lightnings would be our maids.

172. Well, we have found the purest and wisest old blood, and the most spirituate young blood, in company in the jugular veins, and on the way to the expectant heart; and this association of lives continual, not intermittent or divided into pulsations. This seems to raise the consideration of the veins to a new point, and to allow us to say that if the bright red or scarlet blood of the arteries is kingly in its robe, the dark red or purple blood of the veins (in its composure)

is queenly. In its brain-streams it is full of prolification: a royal bed of lives. The physical cause of the colours of the two bloods is not yet conjectured, though the spectroscope has been at work upon it. Swedenborg gives a spiritual reason for the colour of the blood: " It is red by virtue of the correspondence of the heart and the blood with love and the affections. In the spiritual world there are colours of every kind. Red and white are the fundamentals. The colour red there corresponds to love, and the colour white corresponds to wisdom. The colour red corresponds to love there because it derives origin from the fire of the spiritual sun, and the colour white corresponds to wisdom because it derives origin from the light of the same sun. And because of the correspondence of love with the heart, the blood cannot but redden, and point to its origin." (*D. L. W.* n. 380.) Physiology can with difficulty look towards such psychology, but custom and perseverance may do much for it, for the age is one of eye-openings. The Poets precede as usual. Milton says, "Celestial Rosy Red, Love's proper hue." And as regards facial expression, sung and said from the beginning, the healthful fire of love has ruddy cheeks and brilliant eyes, and the sweetness of it also lives in blushes: arteries, capillaries and veins are at their best here. The physiology of expression is always a court of common sense, and this Rosy Red is a witness in it.

Psychology of the Veins.

173. If the arteries correspond to affections for the spirits of the organs to which they give their life, the blood which in its individualities has discharged those affections in the capillaries,—the privities of the conscience

of the affection,—is urged with the purpose and appetency of seeking the heart to recover the life it has given up, and which it cannot but desire to replace. The heart alone can restore it, but with the lungs. The veins are therefore desires, while the arteries are affections, and the heart as love stands between the two. The veins correspond to the pressure or desire for service in the general blood itself. The entire education and fitness of the blood is effected in them. They are essentially masses, not individuals, excepting for purification in the liver and the lungs. For life-seeking they are "the masses." And their sought life consists in the reception of the new infant blood, and in their submission of it to the cardiac and respiratory powers, in which it grows to maturity with each pulse and breath. The veins are as homes which love their own inmates or contents, and shelter them; the arteries are public lives, which affect the forum of the system. Every affection of adult love is twofold, or has two sides, one public, the other private. The man goes forth to his calling which concerns society and its wants, and returns to his own wife and family, who are his belongings, and to his own bed. There is affection in both cases, but the one is a centrifugal, the other a centripetal affection. The one is the love of the morning; the other, the love of the evening and the night. The family of the man's love, at least in the correspondence of the body, is around him, and in them again he seeks his youth. Sleep, the giving up of himself, is the moment of transformation of affection, when it is neither public nor private, but both merged in one common end, in which self-love and social love are the same. Is this the lung-capillary state in which the blood is lost to the general system, and elevated into a higher sphere where airy dreams can come? All these things take place simultaneously and perpetually in the body and the mind.

The Capillaries and the Vasa Vasorum.

174. The representation of the mind or man by every part, is essential in the psychology of the body, and should never be lost sight of. It exists in the capillaries preeminently by constant changes of the blood and the vessels according to the states of the affections. It is signalized in the face by changes of colour; and in old age the skin often reveals these changes to the senses by admonitions and irritations. For the capillaries, as we have said, are the homes of the blood, where it shows its odd private ways. It is known and judged there as nowhere else. The heart is at one end of the blood-

system, the capillaries at the other. They correspond in sympathies. Whatever in large tracts the capillaries feel, the heart feels; and *vice versâ*. Each plays upon and alters the state of the other. So the capillaries are the individuals, and the heart is their State.

175. The capillaries have their life or nutrition only and directly from their own blood. In this they are like the heart, and all their outflow is into veins. This is another proof from that vast objective of the heart,—the capillary system and firmament,[1]—that the coronary vessels are an outflow from the heart, and not from the aorta. For the capillaries want no nutrient

---

[1] "The area of the arterial system increases as its vessels divide. The capacity of the smallest vessels and capillaries is greatest. It follows that the arterial and venous systems, as regards capacity, may be represented by two cones whose apices are at the heart, and whose bases are united in the capillary system."—Quain, Vol. II. p. 184. There are also two cones in the heart itself, and the coronary cone is the greater of the two.

vessels, but are nutrients not only for themselves, but to all the organs. The sustaining blood has the body of life within it, and the quickening nerves give the organ the power to choose it wisely and well ; the lungs supplying ultimate power to attract and take.

176. But there are other vessels still, the *vasa vasorum*, the blood-vessels of the blood-vessels. Of these the following account is given :—" The coats of arteries receive small vessels, both arterial and venous, named *vasa vasorum*, which serve for their nutrition. The little nutrient arteries are not derived immediately from the cavity of the main vessel, but pass into its coats from branches which arise from the artery (or sometimes from a neighbouring artery), at some distance from the point where they are ultimately distributed, and divide into smaller branches within the sheath, and upon the surface of the vessel, before entering the outer coat

where they are distributed. . . . Minute venules return the blood from these nutrient arteries, which, however, they do not closely accompany, and discharge it into the vein or pair of veins which usually run alongside the artery. . . . The coats of the veins are supplied with nutrient vessels, *vasa vasorum*, in the same manner as those of the arteries." (Quain's *Anatomy*, Ninth Edition, 1882, Vol. II. pp. 189, 191.)

177. The arteries are a continued heart. In the heart, however, the arteries and veins are clubbed in a conical organ, while in the body the two sides of the heart are prolonged into separate trees of arteries and veins. Can we then still see in the arteries the same principle of taking the blood for themselves which exists in the massive heart, and of which the coronary vessels of both kinds are the outcome? Four coats or membranes are reckoned to the arteries. Lancisi, one

of Swedenborg's authorities, says of the fourth or innermost coat, "that it is the finest of all the membranes, but strong, and although it contains only the most minute foramina or orifices, it is nevertheless transparent, and may be seen through when held up to the light. This coat of the arteries corresponds to that which lines the inside of the ventricles of the heart, and which being made up of a most dense tissue of villi, alone confines the more subtle and volatile particles, which would otherwise no doubt in great part make their escape." Here there seems to be the same mechanism as in the heart itself, namely, a porous inner surface for taking in a select part of the blood at the moment when the artery embraces and sends forward the volume of it. In this case the artery of itself supplies itself with blood life as in the heart, and especially supplies its own muscular coat. After which the venosity of the arterioles, of which the foramina above mentioned are the orifices, begins. They empty their contents into the arteries which are within the sheath and upon the surface of the vessel and its outer coat, as the coronaries into the aorta. The so-called arterial *vasa vasorum* do not appear from Quain's *Anatomy* to be distributed to the inner coats. The vessels of the foramina supply the link. The same reason may be urged for the blood-supply in the arteries as in the heart; namely, that an artery full of blood, and in which the most spirited blood obeys its attraction to service the most easily, supplies itself, and does not require to be nourished from any more distant source. Swedenborg's statement, that every viscus and organ has its own angiology or science of vessels, requires to be remembered here. Our explanation is proffered to await anatomical research. The question still occurs with regard to the systemic veins, how they fulfil the coronary analogy as exhibited in the right auricle and ventricle of the heart.

Nutrition.

178. The Microscope has contributed greatly to our experimental knowledge of the nature of nutrition. The assimilation of food prepares for it; Food being the material quarry from which our marble is won : but understand by this aërial food also from the mountains of the air. But the brain, the heart, and the lungs are the artists which under nature produce us. We are built not from without as human artists build, yet with an omniprevalent idea of the whole man, as good artists work with a unitary conception and imagination. So the repair of the body, continually needed in the waste caused by function and exercise,

and its increment wanted in growth, and its newness for health, are done in wholes and not otherwise in particulars. It would be difficult to say how much is gone and how much replaced every morning. The learned estimate that in seven years we have an entirely new suit of mortal clothes.

179. All this goes on life-wise. The blood becomes the organs. The first stages are cellular, — cells, themselves already organs. Once in us, nature makes obeisance, and there is no external gap between blood and organization. Minute, even infinitesimal wholeness prevails. In that wholeness soul-formation is represented. Absorption of old individua,—human atoms,—takes place with even pace on the same lines as deposit. We may think of it as a system of dissolving sculptures, and nutrition as a new field of apparitional sculptures. Entire forms, infinitesimal though they be, disappear, and

again loom up, as the mind gazes at these living studios. You see the expressive face of order and freedom continually made in circumstantial detail with endless difference for each place and part in the integral man.

180. There is no need to dwell again on our oft-told tale. "The Soul is form and doth the body make." That is the sum of it. The mind, the brain-master, gives the secondary life of it. The heart and lungs, the representative mind, give the blood and breath of it. So every day the old man and the new man are self-contained, and fenced from without into psychological freedom, and the old man is continually fading out of use, and the new man punctually being placed in his stead. A most ghostly and most bodily theatre: merely the union and play of the soul and the body over again.

181. The soul builds the whole body at

first from the divine plan contained in itself out of the storehouse of the mother's heart and womb-life, in the sensible image at the bottom of the hereditary nature of the father. It is all drawn from without, and converted within. The faculties are formed into power one after another, beginning with the senses. Heat and light are lifted up by the descending soul, and thus and then become impregnated with recipient forms of affectional and mental ensoulments. In the formed body the above is the model and way of all nutrition. In the leasts the soul-work repeats the great plasticity of the whole. Building, rebuilding and unbuilding are the same in end; the unbuilding is but the defeat of death in his first graveyard, or rather on his first funeral pyre, by the transference of mortal into quasi-immortal memorial states, the then imperishable properties of the soul, and the registers of its life led in the body. Affection and thought seem also to perish, and the will and understanding to be subject to oblivion; but they are in their garnering the substance over which the ego, the person, presides for ever.

How the Organs feed.

182. A doctrine to be reiterated is, that the heart's pulse carries the blood only to the threshold of its places of work, the organs, to be taken by their expansion into the capillaries. Like infants they all suck the teats of their mother, the heart; but the milk of its love is nowise forced upon them. In this way all secretion and excretion has its own animus with it according to the choice and analytic faculty of the nature of the organ. Otherwise the organs

and the body itself would be a mere ooze or delta, dead and tideless; and the laws of membranous filtration called exosmose and endosmose would supplant the ratios of bodily intelligence or lung-life. For the lungs are at the outside of all, and apply the organs to their native blood. As persons the organs freely feed, — their arterial gift; they put by maintenance into appetite,—their venous function; and they open themselves to eliminate,—their excretory and rejectory function. All by their own animus, spirit or breathing.

183. Dr. W. B. Carpenter invented the phrase, "unconscious cerebration." It means actions throughout the body directed or done by automatic mind, for cerebration or brain-work amounts to mind and its processes. His thought covers and includes all the affections, intellections and senses of the person, for all these are brain-work.

but in a new series under automatic or in fact cerebellar conditions. His phrase might have been cerebellation, which would have then included the word unconscious. Thus this great investigator, perhaps himself by unconscious cerebration, transplants will and mind into all the organs and viscera of the frame, and into their performance of their functions.

Arrested Physiology.

184. The course of our subject asks the question whether it is possible that vast accumulation of facts and observations in the department of anatomy is compatible with absence of the knowledges which begin to constitute proper science here, and with knowledges higher still which gain a position for the subject in the human intellect? Such a condition does

seem possible. Sensual light may exist in great apparent brightness apart from intellectual light, and if it is ardently loved it may breed a limitaneous complacency which puts out the superior light. The extinction is perhaps unsuspected, and the process of ignoring the higher light whenever it threatens to come forth becomes automatic. The age abounds in mental extinguishers; mere heads for death, but quick to put their own grinning skulls over the lights of heaven. Organic science, hugely sensual, is in a dark age without knowing it. What are dark ages? They are those in which mere learning, whether from books, or from the volume of nature, is prosecuted by specialists who are accepted as the last authorities, and carry their readings and facts to no superior tribunal. Nature becomes a fusty and worm-eaten volume when you hoard its contents for no ends above or better than nature. This you do if you eliminate the soul of man and its interests from the pursuit. Also if you treat the body as if the soul were alien to it, and as if your self-issuing intelligence and its inventions had plenary property there. Both these ways destroy that regeneration of knowledge which alone is true progress. Surgery, the larger and more reckless it grows, can be more and more immersed in those dark ages by dark deeds. Experiment and Chemistry are not lights that preclude dark ages; but mind, conscience, soul, religion, the Lord.

185. A more sensual discovery well and newly made out, a most needed discovery, may by its lower interest extinguish an ancient light that ought to be added to the new light and kept burning within it as a golden candle in a then golden candlestick. The discovery of the systemic

circulation by Harvey, Malpighi and others, diverted attention and belief from the cardiac psychology, which, when followed out, led to the doctrine of the proper circulation of the heart. This doctrine however, though delivered long ago, has no place in modern thought. The strong sensual light of the lower doctrine, true and valid in itself, has interposed a cataract of comfort before the higher vision.

186. The self-satisfaction of boundless observations which no one gainsays, keeps important discoveries fixed, and prevents investigation into their secrets; for every true and especially every large discovery is capable of being carried upwards and inwards as well as forward and outwards. So it is that great problems are deserted and infinitesimal particulars are accumulated. We may say broadly that there are no perceptions at present of man's body, but

only sensations of the body of man. These registered,—ably registered indeed,—are our physiology. The mind revenges itself for its extinction by the will. Made small,—into a brainy heap of microbes,—it is drunk with the microscope, which is its perpetual consolatory dram; truly a glass too much for it.

187. The quest of *minima* in itself is as legitimate as astronomy resolving nebulous patches into distinct stars. Yet the exclusiveness of corporeal star-gazing carries with it effacement of the substantial parts of knowledge. It makes the microscope overwhelm and despise the naked eye, and common sense and common observation suspend that breath which is the abode of the spirit of intelligent and practical discovery. It dilutes our perceptions of uses, ends, causes and functions, which are the heart and lungs of our knowledge. In no other walk can it do so much mischief.

Chemistry of right deals with atoms and compositions and decompositions, and thereby attains endless new compounds. We may say, since Dalton, that atom-volumes are its organs. It is also dead from beginning to end, which is of humane as well as rational importance when breaking and sundering are needful. Organism is otherwise given, and is from the first alive, and does not bear destructive analysis, but is cruelly defeated and un-facted by it. The main business of physiology, and above all of psychology, is to see the organs *in their places*, to see them as alive and at work, to demand their totality, and always as a functional part of the whole man,—body, soul, and spirit. His biggest organs are his atoms: in reason they are no fields for cutting up. Keeping this steadfastly in view, sufficient light may come out of it to the retired sphere of leasts; for in man the leasts are the greatest and the total over again, but in the shroud of intimate nature. Thus much is rationally certain; but the bodily eye cannot see it; and when the microscope gains the upper hand, it makes the mind's eye deny it.

IV.

Psychological Notes.

Revealed Psychology.

188. The Word of God so accords with the general, perhaps universal, testimony of language in regard to the organs of the body as mental designations, that we may affirm that Revelation is itself the source and sanction of this hieroglyphical usage. Anatomy and physiology are not the origin of it, and at present impugn it. Deriving it from the Word, we there learn by immediate inference that the heart stands for the embodiment of love, for we are commanded to love the Lord with all our heart. We can love only with the organ of love, with that in us which

loves; from our own heart, will and centre of love.

189. The lungs are not mentioned in Scripture, but instead of them the breath, the spirit. This was breathed into man's nostrils in his creation; it is called the breath of lives; "and man became a living soul." The nostrils signify the first general perception of life, and belong both to smell, which answers to perception, and to respiration, which answers to thought about what is perceived. This high in-breathing corresponds therefore to spiritual life; the life of love to the Lord and the neighbour commanded in the first great commandment, and in the second which is like unto it.

190. The head at the summit is named for wisdom and its rule; and where wisdom is, there is love: and where wisdom is not, the whole head is sick and the whole heart is faint. Kings in their anointing were made

by the head, and Baptism or the first seal of spiritual life is by the same. Brotherhood also. "Behold how good and how pleasant it is for brethren to dwell together in unity. It is like the precious ointment upon the head, that ran down upon the beard, even Aaron's beard, that went down to the skirts of his garments." It corresponds to unity, and draws all things to itself, as the heart pours forth all things from itself. The head, the heart and the breath are constantly mentioned in the Word, and with uniform signification: in the Lord and the regenerate man in a good sense: in the opposite sense where evil is treated of. The latter sense may mark any objects; as when it is said in the Psalms, "The Sun shall not smite thee by day, or the Moon by night," in which text Sun and Moon are clearly evil things from which the Lord will keep us. So also when He says, "The Heart is deceitful above all things and desperately wicked," this is a heart opposite to that with the whole of which we are commanded to love Him. The explanation may be needed on account of the opposite senses of love and self-love as used in the present essay.

191. The face in the same way as the heart is used in the Word for the interiors of the mind which the face reveals: seeking the Lord's face, we convert ourselves to know His revealed will instead of our own, and to be converted to accepting it. All the senses located in the head are psychological terms: the eye is the organ of manifest truth; sight is the discernment of it; and vision the elevation of it. The ear is obedience, the organ for hearing, listening, and hearkening to the words of wisdom. The nose is perception, and examination of qualities by perception. The taste, sapor, tests by sapience what should be received

P

by the mind, and what rejected. And the tongue, a double organ, declares by its vocal connexion with the understanding—the lungs—the decision of the testing. The tongue with the mouth is a world of speeches as the ambient air is a world. The sense of touch covering and embracing every sense is the whole human form as sentient; the heart of the outside; we touch and are touched according to our affections; and love in its mystic ways and groves haunts the hands of touch. Intimacy and power are both signified by the touch of the hands. "Touch Me not, for I am not yet ascended to My Father and your Father, to My God and your God," signifies that He cannot be intimately known for the One only God until the last ascent has been accomplished. The subject of this sense is immense, since to touch is eminently to feel: and all feeling from the innermost soul to the body is expressed by it both in Scripture and in common discourse.

192. Proceeding downwards, the neck is mediation, known perhaps chiefly in the Word on the obstinate side. Stiffneckedness and hardneckedness are psychological equivalents to refusals to allow the wisdom of the head to pass into and influence the breast and the heart. All the body comes down through the neck, and when its mind is disobedient and unyielding, wisdom is unrepresented, and unregenerate nature prevails: the influx of the highest things into the lowest, and the conjunction between the two, is stopped: goodness and truth are intercepted and rejected. The neck is yielding when it bends to conscience and divine command.

193. The neck, however, as a human ideal, I will not say scripturally, speaks in sculptures of correspondences. Love in the power of

stateliness has its palace there. Falling upon the neck is an act of ecstatic devotion, of the tenderest union of the soul. Love puts its pure arms, its powers, round the neck of the beloved. The Soul and the body meet in the neck and are there conjoined. Erectness and uprightness are in it as the manifest pillars of their temple. Holding up the head, the human attribute, is the claim of innocence, the candid pride of its humility. Between the grave head and the burly body the fragile narrowness of the neck pleads for delicacy of affection in the grain of its sensitive skin. Beauty speaks in it not as in the face by mental expression, but as by ineffable endearments that make no claims. For it is an organ of transit carrying its treasury of lives and parts up and down not for itself. An ascending and descending staircase and ladder of mysterious dignities and powers. It is a lively type of influx between separate planes, and of correspondence itself; of the union of worlds ; a Mediator between them.

194. The bosom is love extant in character, and receiving love, all through the Word. Lazarus was carried by angels into Abraham's bosom. John, the beloved disciple, lay on the breast of the Lord. Hearts are opened when the breast is opened. The ribs build the love as its chamber: dear to the heart: and they are built continually into all its thoughts through its moving marriage with the lungs, and give the mind bone and muscle: may we poetize or fabricate in saying that Eve, the primal mother, still has memories from their foundations? Think too of the heart and the reins as constant associates in the Word. In this context they are lower and higher consciences, united, — searching and purifying. The bowels are yearnings,—constant divine names

for mercy and pity. The womb lives in these heavens as the seat of the love of infants: the Lord takes the man out of his mother's bowels to transplant him early from creation into redemption. May we not then say to some ears that in the Word, touching the organs and members of the human Form, there is a Revealed Psychology? The echo of it sounds through the ages, and is the gold of their Poem. Love is tuneful to it. There is no speech or language where its voice is not heard.

195. The gait, bearing and postures of the body also are used interchangeably for spiritual states. Many instances will occur to the diligent reader of the Bible. Uprightness is one that prominently occurs. I decorate my pages here with a physiological truth, which appertains not only to the head, the body and the limbs, but to all the organs *in their places.* " It is apparent

that throughout the animal kingdom the gift of physical perpendicularity has been reserved to man alone; that by and through it man is installed in his recognised lordship over the beast of the field, and that in it we may discern the boundary between the soul of the beast, which goeth downward, and that of man, which goeth upward. It is further apparent that physical uprightness is both significative and comprehensive of moral uprightness also; is divine in its institution, humanizing in its working, and the insignia royal of the great king, that man is born heir to heavenly immortality. And so irrepressible was the perception of this significative nature of animal erectness, that Grecian philology baptized man AN-THRŌPOS, the being with the upturned (heavenward) face."[1]

[1] A Rational Treatise on the Trunkal Muscles, p. 3. By E. P. Banning, M.D. New York, 1868. This work is an important contribution to an integral physiology, replete also with psycho-

196. The feet as well as the hands are the ultimates of spiritual life. The ways of life are in the feet, the acts of the ways are in the hands. In the Word the risen Christ sits at the right hand of God: He has all power in heaven and on earth. He has put all things under His feet: He has subjugated Hell and redeemed mankind. These are divine human attributes made accessible to every mind in the incarnation. They are the *Divinum Naturale* and the *Divinum Sensuale*. They justify the position of a divine psychology, — of a God-Man. All other organs, expressly or by implication, are added to them in Revelation. On the finite side the hands and the feet take up

logical insight, and of practical value throughout the domain of medicine and surgery. Our conceptions of the power of the lungs upon the body are amplified by it. And it shows how the trunkal muscles united everywhere to and in respiratory acts, keep the organs of the body to their positions, and prevent them from dislocation; from limping and hobbling in their lives and functions. See my "Medical Specialism" in the *Homœopathic World*, September 1889.

the representation. We are judged by the works of our hands. If our feet are clean, we are clean every whit. For the feet here represent the sensual natural man in whom the mind ends, and when he is regenerated the whole mind is saved. " How beautiful on the mountains are the feet of those who preach glad tidings!" Works and ways are the actual life without which internal life has no outcome and no basis. They are *ultima ratio rerum;* the final stamp of the will.

197. The blood too is a spiritual symbol throughout the Word. The Lord's blood is shed for our salvation; it is His divine truth given afresh in the combats of His life, and to be lived in us in the battles of regeneration. The Lord's blood is divine truth as His flesh is divine good, and we are commanded to eat His flesh and drink His blood, as meat indeed and drink indeed. Blood also is the symbol of all kinship: God has made of one blood

all the families of the earth. His blood is the bond of all love, and the redemption of it: the heart of heart. In the highest sense the blood is the life; the life is in the blood. The spiritual essence of race resides in it, and man's capacity of elevation and amenableness to divine instruction is determined by it; for in a deep ground it is the will of will, even freewill itself, the determinations of which are the man's proper nature.

198. The consistent usage of the Word in regard to the organs of the body, and the language of mankind so far as it accords with that usage, enshrine divine correspondences for our instruction. These are actual ultimate facts of the human form, and reveal its soul. They stand above speculation, and reverent thought flows natively into them. They are revelations of organic knowledge; and will come down into the true ends and mental uses of physiology as

its ordaining and administering principles giving it spiritual life.

199. In sum, the inspired Word from beginning to end reveals in all detail that there is a divine humanity, one God in one Person. This is a mental fact, but involves the human form. It is an infinite human form, but finitely presented from the first in Jehovah, and since the incarnation realized in the Lord. In heaven as an appearance it is veiled in its own glory, and is seen from afar as the Divine Sun. But because the Divine Humanity in itself is ineffable and unapproachable, and infinitely above the heavens, it comes to the lowliest Christian as to the highest angel in the Word in its natural and spiritual senses, and the Lord Jesus Christ has His perpetual appointment with us there, and meets us as God-Man to Man.

200. We quote from Swedenborg: "Every

one who believes that God is a Man is able to affirm for himself that there are infinite things in God. For because He is a Man, He has a body, and all things belonging to the body, . . . which things taken together make a man to be a man. In created man these things are many, and regarded in their contextures are innumerable; but in God-Man they are infinite, and nothing is wanting, and therefore to Him belongs infinite perfection. A comparison is made between Uncreate Man Who is God, and created man, because God is a Man, and it is said by Him that the man of this world was created after His image and into His likeness" (Gen. i. 26, 27). (*D. L. W.* n. 18.)

201. "That Infinite things are distinctly one in God-Man, may also be seen as in a mirror from man. In man there are many and innumerable things, as was said above; but still the man feels these things as one

thing. By sense he does not know anything of his brains, of his heart and lungs, of his liver, spleen and pancreas; or of the innumerable things in his eyes, ears, tongue, stomach, generative organs, and the rest of his interiors; and because he knows nothing of these things by sense, he is to himself as one person. The cause is, that all these things are in such a form, that not one of them can be absent; for it is a form recipient of life from God-Man, as was shown above (nos. 4, 5, 6). Out of the order and connection of all things in such a form, a sense, and thence an idea, is presented, as if there were not many and innumerable things, but as it were one thing. From this it may be concluded, that the many and innumerable things which make as it were one, in man, in Very Man, who is God, are distinctly, yea, most distinctly, one." (*Ibid.* n. 22.)

202. This extract contains a warning not

to think naturally and sensually here. The infinitudes in God-Man preclude it: they are to be affirmed with all the intellect as "distinctly one," and in prayer and praise, which alone can approach His throne, no separation is allowable. Whosoever values his own mind will rest in God as his Father in the heavens, and will reject as dangerous all thoughts outlying and thus infringing the Divine unity.

A Word from Greek Philosophy—Clinamen.

203. Once or twice the word Clinamen occurs in these pages, and the explanation of it demands a few words. It is a metaphysical notion brought down by Epicurus and illustrated by Lucretius, and has arrested the attention of thinkers to the present time. A striking conceit, it looks like a survival of some spiritual perception of which it was the

last breath. The following statement may give the reader some notion of it.

204. The Epicurean view of nature was that atoms are the original beginning and sum and substance of all things, and that void space is their infinite playground from eternal time. In said time as they hurried on incessantly, they underwent, tried, experienced, all possible combinations and happenings—all cases—as it were dicers in universal chance; and when anything was made or occurred by said chance, such thing maintained itself as best it could, and became a field of its own laws. If it was seed with soil to it, it behaved as a seed and grew. When the primordial atoms had diced forth a man, he stood his ground, and if woman were accidentally added, a race was initiated. Of course any number of stages and platforms would occur before these circumstances were gained. But thus, in time, sun, moon

and stars arose, and a Cosmos, in the midst of an infinite atomic accident-universe. And atomic time being eternal, there was an infinite chance that whatever could come of the moving atoms would some day arise. The world was a game in which in the long-run everything we now see and are drew the prize of existence. Also, as experiment of atoms was constant, there was an infinite chance that everything would some day be resolved into them, and lose its foothold. This play of construction and destruction, life and death, would be endless. The immortal thing, however, is death, *mors immortalis;* all atoms, however, surviving.

205. Here are no gods or mental originals, but the blind rush of everlasting nature is the lord of the presumed infinite situation. Epicurus, however, observed that in the scheme of atoms descending straight from eternity of time into infinitude of space, there

was no chance,—and chance for everything extant was postulated,—of freewill for man; and freewill was a dogma which Epicurus strongly asserted. For this faculty to exist, something was wanting which could break the chain of necessity, *quod rumpat fœdera Fati*. He found the missing link in kinēsis kata paregklisin, translated by Lucretius, *clinamen*, leaning; but it imports motion according to mood or inclination. This was really a psychological induction from the assumption of freewill reacting against the dogma of *plagæ* or blows of atoms striking each other from without; for evidently the self-centre called *libera voluntas*, freewill, could not arise from such cudgellings. In other words, Epicurus slid inclination in as a little wedge, and ultimately as a way of accounting for freewill. The atoms themselves are infinite in number, indivisible and indestructible, and have no sensible qualities.

As such, could they approach each other, they would be always either parallel, or at angles, and the parents of nothing but distances, or collisions. But while travelling down space in straight lines, "at quite uncertain times and places they swerve from the perpendicular to an imperceptible amount." Out of this swerving which was "so small that it could be no smaller," he confirmed his assumption of freewill, which called in his mind for the swerving.

206. It requires a skilled metaphysical hair-splitter to see the great consequence of Freewill in this most minute indirection. The first stage is the leaning,—the native leaning,—of the backbone of atomic nature itself: "c'est le premier pas qui coute." In this first stage the chain of fate is broken: the law of universal necessity has disappeared: some atoms have set it at nought; and where a thing once under despotic fate

is emancipated, that thing is in freedom ; it is at its own bid. The second stage is that the arch atoms leaning to each other, touch, unite, conjoin, couple ; and in this way compound ; and produce secondary and tertiary products, and any number of more remote edifices and structures of form. Lucretius applies this not only to explain freewill in man, but to account for the origination of motion in animals ; for instance, to the faculty by which horses guide their movements, start, run, and stop. But the leaning takes place only in certain atoms ; perhaps because if it occurred in all, the universe would be stipated with body, and flesh would be the rule and tendency of everything. Perhaps also because this would cancel the atomic universe, and plant upon it a universal freewill.

207. The *clinamen* or leaning is therefore a part of the original possibilities of the

infinite atoms. Are there then two kinds of atoms, of which the one inclines to society with its fellows, while the other remains in singleness? Or is there an agreement by which the one set compacts with the other about the destinies of each? In either case we come to mind, to inclination, desire, love, in the atoms; to a mutual recognition and appreciation; very near to a cosmical heart and intellect as their sum. Atoms with such a sidling and slope and attraction and final adhesion to their fellows, are clearly prophecies of wills and understandings, of men and women, in fact of Darby and Joan who are their ultimate outcome.

208. Worthless in itself, a mystery of thought hovers over this little quaver and tremble of the atheistic mind; an apparition of *Deus maximus in minimis*. The reader will easily divine our reason for making a note of it. For it recalls our bronchial

arteries which enable the lungs to swerve aside into an identity of their own; into an independence from the rush of the heart, by an emissary remotely born of the heart itself; and thus to assert the breath of freedom for the whole man. The groundlessness of the dogma is that it is a mental fact, nay a potency towards all mental faculties, weltering in space without any containing body or mind. No Deity directs it. Its value lies in its suggestiveness. In this regard it has a still small voice, and preaches of the legion of particular events consequent upon the smallest deviations and opportunities; and of differences and vital states in their minutest initiaments growing evermore from the round bole of the spiral tree of inclines. I cannot but imagine that this doctrine has lain deep in the mind of physicists thinking towards causes, and that it may be a precedent factor in the arcana of the evolutionist hypothesis. On the other hand, as introducing a new power into nature without a cause though with a strong human craving for one, it admits a being uncaused, which is a boundless being, and commanding, nay omnipotent, because having no antecedents which command it; with the result, as the event postulates, of immense creativeness. This cause can be none other than an eternal God; and the leaning of nature wherever it occurs, uncaused otherwise, can be none other than the work of the finger of God.

Matter and Life.

209. Matter does not enter into organization,—even into vegetable forms,—without putting off its names and qualities, and taking on higher rank. In all such cases it "suffers a" life-"change into something new and strange." The names of external sciences

exist only by courtesy and convenience, and as waiting for new names, in the human body. The air comes into its lungs, and is breathed out again, but as air it is never inside us : it is in ample reservoirs of its own, but still, in their most secret places, it is on the outside of the tissues, or communicates by the mouth and nostrils with the open air. The food and drink taken in by the mouth, as food, are on the outside all the way from the lips downwards. So soon as the air touches the blood, and receives its modification, a biological fact, a life-fact not chemistry, has been transacted, and the nameless result belongs to arterial blood. The blood, ensouled, brings this about by touching and magnetizing the air. The Soul takes it to the air for that purpose. The blood sunders the air into horse and mule ; rides back as a knight into the heart on the nobler animal ; and sends away its old age into nature on the mule. But though the air be a subject of chemistry, and can be decomposed into oxygen and nitrogen, there is no chemical action in the body but what is vital chemical. So again the food on entering the body departs from chemistry into the needs or digestions of life. Until some digestion takes place, it is outside of us in the stomach although in a different sense to the "outside" when it is on the dinner-plate. In the latter case it is offered for service to the man: in the former it has been received for examination by the stomach. But bothwise, it is an external body until assimilated into blood and chyle. Assimilation to a man's body is vitally human, not chemical.

210. So also the light of the sun which feeds vision, in the eye and the brain-eye, and in the eye of the registrations of memory, meets the descending intellect, which consumes it into understanding, or living light.

211. Whensoever a higher factor actively touches a lower in the same organism, it acts by sight and insight, by spiritual selection: if it can adopt the whole of it as suitable to itself, it works with that whole as an assistant, and reposes upon it as a foundation. If it cannot do this, it severs the lower plane, discriminates what is suitable in it, and rejects the remainder or remainders in order and series into appropriate places. This takes place throughout our bodies. It is a correspondent of the action of conscience in our lives and of intellect in our situations. But the higher does not accept the names and ways of the lower things as binding on what it does with them. These designations perish, as the chemistry of the air on being touched by the blood perishes in the presence of the magic and mystery of the heart.

212. Also whenever any elements in the

body or the blood are effete, in the degree of their worthlessness to final causes, they are out of the system, or on the way out. They would be chemical if they were not suspended by life as privileged : but were they chemical, they would be destructive. The more they can assert chemical forces, the more they eat away organic parts, as in cancers and other malignant diseases. Yet they never do this wholly unless in the corpse; which is an immediate field for the chemistry of nature.

213. The human body lives in a world, and for its senses and upper faculties wants an immense cosmos for its house. The world comes to it, but not into it without changing its nature. There is no physical influx. The blood divides the air, and is victorious over its physical expanse. But what takes place is yet unrevealed; and what is called atmospheric chemistry stands

in a gap of vital knowledge. We note the change from venous to arterial blood in the capillaries of the lungs : and we know that neither oxygen nor nitrogen are parts of life ; and therefore that they are not present as such in living blood. We know that they are ministers to the blood, and that is all we know.

214. The chemical analysis of the air, valuable as science of matter, does not help physiology, still less psychology, unless spiritual light be added. Ordinary thought can work with air unbroken as well as with oxygen and nitrogen. Taking air as it is, you know that it is the matter of breath. You know that flame burns in it, and that blood requires it. You know that its pressure opens the chest and the whole body from head to foot by the attraction of the lungs. You also know that the direct chemistry of it ceases with the chest.

215. A thought for medicine comes out of the fact that our bodies are islands in the great world with all its commerce for their use, but that nothing lands in us but by our own organic choice, and at our boundaries. Neither heat nor light, nor ether nor air, nor material food, is allowed to enter, without being received organically, and becoming constitutional in us. The body is talismanic in transformations; miraculous every moment. Intrusion is only by violent mechanical agents, or by poisons. Therefore strong drugs and erosive substances as a part of medication, are hostile to the fundamental nature of organization, and affront recovery of health. Therefore also when these are given in infinitesimal doses, and as compliants with disease, they have rights of entry where there is a breach, and upon this rock Homœopathy is founded.

216. Such infinitesimal agents, as *similia*

to disease, belong to it at the lower end,
where they are harmless, though of drug-
nature, and generally of poisonous nature;
at the upper end they belong to the very
prudence of poison, which opposed by skill
to the poison of disease,—always tending to
destructive chemistry and its deadliness,—
amounts to reaction against the disease,—to
a clinamen or mood destructive of it, and
turning its front. We must remember that
in health a quasi-mental skill reigns in all the
organs. In disease, a corresponding defec-
tion. Yet, when the organs are ill, they
crave to be well. Half-shipwrecked, they
know when the life-boat is coming. The
impaired skill within calls for skill to be
supplied from without to awaken it into
activity. In this way there is a natural ratio
between the New Art of Medicine and the
wants of suffering humanity. *Similia simili-
bus curantur* is a law of cure that applies to

the body as well as to the mind. It is the function of *the bronchial arteries* in both cases. Disease is the selfishness of separate parts enforced against the interest of the whole man. This is met by turning the selfishness upon itself, and causing it to rebel against its own consequences. All but final and mortal diseases and evils can best be treated in this way; and on the bodily side even these can be soothed and not irritated in the inevitable.

217. The idea that everything in man is mere Chemistry is well-nigh formulated into doctrine, and that life is an upper result of chemistry. And with it we are bidden to be fed considerably on chemicals; while bread and wine seem to be waiting their doom from the creative laboratories. So great is the revolution that chemical analysis has wrought in physiological speculation and medical practice.

218. Let us, however, pause to concede that there are two chemistries: the chemistry of analysis: the splendid external science of to-day. It is always a reduction to the mineral or dead elements even where it is called organic chemistry. The other kind is organic chemistry proper if you please so to term it, and consists of the products of those living alchemists, the organs themselves. Organic chemistry decomposes and recomposes only in the living body. It seems to be of pantoplastic endowment, as Swedenborg describes the blood to be. " The differentiation of cells," says Gray in his ANATOMY, "is a term used to describe that unknown power or tendency impressed on cells, which to all methods of examination now known, seem absolutely identical, whereby they grow into different forms; so that (to take the first instance which occurs in the growth of the embryo) the indifferent cells of

the vascular area are differentiated, some of them into blood-globules, others into the solid tissue which forms the blood-vessels" (p. cxv., Ed. 1883).

A Door to Psychological Chemistry.

219. The lungs correspond to the intellect. "The air corresponds to all things of perception and thought." Whatever so touches the human body as to be conjoined with it, becomes a subject of physiology, and also passes over into psychology. The body knows it first, and then the soul knows it. The lungs are constant and craving appetites for the air because it is ordained for bodily and psychical union with them. In other words, it corresponds to them, and feeds them, and they feed upon it, by a mutual natural correspondence.

220. Each unit or individuality of the

blood into which it is reduced as a single personality in the capillaries of the lungs, rolls quite away from the heart into the embraces of the new movement of respiration, and itself is compelled to breathe. All living motion in the body is new life. The lung-motion, the breath, of each globule of blood, is itself a universal factor of life, and like wheels set under every other function, prepares the chariots of the body for rapidity in uses and ends. But the breath, the motion, breathes something, to wit, the external air, and makes it into internal or living air. It is not only the lung and air cell but the blood-globule which is breathing. In the lung that globule is a lung. We cannot see its mouth or trachea, or its ribs, but the function shows that the whole lung-form and power is in it, and that it takes in and gives out the air. This air chemically consists in the present phrase of oxygen and nitrogen. The air is passive, intruded by the pressure of its column. The blood-globules are actively alive, and in their motion seize the passive air. The spirit in them takes the oxygen, the spirit of the air; the body of them takes the nitrogen, the body or flesh of the air. Whatever of either is not required, is breathed away from the mouth, and carries along with it the impurities of the lungs, of their blood, and of the nasal and buccal passages.

221. We may therefore say that the oxygen is the spirit of the external air, and the nitrogen the body of it. Perhaps their union in air is not a chemic but an epichemic conjunction, the one functionally, not locally, superposed upon the other with an inductive or correspondential influx into it: a free marriage bond, but not a loss of either in a *tertium quid.* It is as the case of the heart and the lungs over again: the heart being

flesh and the lungs spirit. Certainly the integral air gives function-play to the lungs, and this, by its whole weight. The oxygen externally inspires the blood, or the blood respires it. As certainly the nitrogen is a fleshly-bodily element, is the very muscle of blood, and in nitrogenous food is used by the heart-muscle for motives and by all the muscles for motions. So what is body or flesh, breath or spirit, in the human frame, we may equate as spirit and body in mutual relation in the aërial universe. For humanity in form is the exemplar of creation.

222. Already chemistry is a hundred-handed science of Uses. When the psychological door and eye is opened into it, which will first take place from the soul in the human body, and from the correspondence of psychical with physical substances according to the uses of the

lower to the higher, chymic thought will be born, and will give intellectual life new servants, even as material chemistry endows our houses with a harvest of unimagined lights and heats and conveniences.

223. There is a third element in the air consisting of the exhalations which issue constantly from all natural substances, mineral, vegetable and animal. By virtue of this collection of spheres the air is a rich and most complex medium. The oxygen corresponds to thought, and the nitrogen to affection; but the spheres in the air correspond to perception. Were our senses fine enough, the several spheres would be sensed as odours. As it is, they correspond to the nose which is the highest mountain of the lungs, and the most gross among them are perceived on its altitudes. They give the blood life from a higher sphere, and begin in

it all the elections of the veins, and all the aversions of the arteries. Perception means this,—continual approval and continual rejection. In this way all appetency and all secretion is founded in the individual bloodglobules; the air cells being the supreme absorbent stomachs and purificatory organs of the body: the heights of feeding and eliminating. This spheral food is taken in and absorbed, and completes the blood as a microcosm in the microcosm.

224. The whole planet evidently also has a sphere, and a universal and powerful sphere. We see it to-day in a great Epidemic covering our world. It is none other than an evil. exhalation from mother earth herself, and comes from no mundane source less fundamental.

225. The air is named psychologically for both affection, thought and perception. Inventions and discoveries are "in the air,"

when from a common zest and zeal they occur simultaneously in distant persons. The spirit or breath of the time calls forth its own works and words. The sphere of humanity is one, and all its organs livingly intercommunicate: they are members one of another.

The Vital Principle.

226. God alone is life, and God gives life. How He gives it is a question that underlies psychology. He does not create life, for it IS, and cannot be created or made. He gives it to forms which He has prepared to receive it. The accommodation of these forms in various degrees to the one divine life is the only vital principle in fact and in knowledge. Human life is that of which we now speak. The explanation of this, however, applies also, but variously, to all living creatures.

227. Men are receptacles or "recipient forms" of life. How should a form not of itself living, receive life? Only by a divine creation of forms corresponding to life. We illustrate this by analogies. The mind and the brain, though not mere life but life through organism, are pre-eminently alive. The face is prepared by creation to be the recipient and mirror of their manifold states. It becomes alive by communication with these internals; by corresponding with them and to them, and then representing them. It cannot represent them as they are, but only according to the virtue and capacity of its candour; for something is lost every time a superior plane is dramatized in an inferior one. But the expressive face shows how life is received, transferred, and propagated. And the common sense of mankind vociferates that the transaction is a case of Correspondence. The mind is the letter despatched

from a distance ; the face is its delivery. The face lives by this use and service, as likewise does the voice.

228. The hands, too, are alive by their created capacity of representing and corresponding to voluntary action. A man's will is known by the works of his hands. You see his will in them because they dramatize it or do it. He feels it in them because they carry out his conscious purposes. So he, the person most concerned, knows that they are his, and correspond to himself. His feet are in the same case : created engines for any standing which he decrees and any walking which he desires. It is the Form in all these cases, the organic structural form, through which the correspondence is effective. Defects of form prove the same thing by partial non-representation.

229. Forms of life are therefore the account of life, and the life is according to the

form. This is plain from the three prime manifestations of humanity, the face, the hands and the feet, all of which are the graphic exhibitions and telegraphic stations and ends of interior correspondences.

230. The governance is that all things which in their intimate, middle and external or ultimate forms correspond to life are themselves alive so long as the correspondence is maintained. When the centre of it is broken, the man and his organs die out of that world of correspondences. When his heart and lungs fail to represent his will and understanding, he quits this scene, and ceases to live on the stage of nature.

231. But also remember that order reigns supreme in the human form, and that the heights of the organism are in one sense the most alive ; that in this way the cortical brain consists of vital principles bodily, to which the rest of the system stands in the relation

of vital principiates. So each plane of the human form is a universal vital principle to the regions below it. Find the highest form in any series here, and you find out the secondary creative life of the rest which is in devolution from it.

232. This is a plain account of what life is, anatomically, biologically and psychologically, and no other can be given. Life and death each attest that the capacity of dramatizing mind, — conscious mind and unconscious,—is the essence of our vitality. This essence is embodied in structural fluids and solids, all embodiment being dramatic correspondence. So the internal viscera and their organic and secretory processes are nothing but so many fitnesses, or answers and *adsums*, whereby the parts are able to stand up in Anthrōpos, Man, speak for him, and furnish forth his schools. Their forms are these fitnesses.

233. Head, face, eyes, ears, hands, feet, erectness, are thus the mouthpieces and masterpieces of a correspondence omnipresent in the body: and from that centre they proclaim that the inferior kingdoms of nature are continuous theatres of the same cosmos or fitness. But is anything still wanting to the definition of Life? The scientific man can never receive the doctrine that adequate form inevitably houses life, and is life, responsible life, for us; unless he yields to the teaching that there is no such separate *thing* as life, but that Life is the divine being alone, and that all our life and all other life is a gift of seeming or consciousness, constant and imperishable, dependent upon the punctual Uses of created forms to their souls. Both God and the Soul must therefore be affirmed.

234. It would be a relief to the Biologist, would save him fruitless exertion and animate

him to the direct study of central forms, mental and bodily, if he would know that his Maker is his life, and that this life is renewed to him by the Creator, moment by moment, through the supreme fate of Correspondence. Even the heathen poet was moved to say,

"Vita mancipio nulli datur, omnibus usu."

Life is given to no man as a property, to all men as a function.

The Elevation of Faculties.

235. When we speak of the elevation of the understanding above the will, and of the understanding corresponding to the lungs and the heart to the will, the question *naturally* occurs to us how a bodily organ, moored of necessity to its place, can represent such motion of elevation? Are the interior spiritual organs not fixed in a manner

corresponding to those in the body? The answer is, that in both cases it is the use which we make of the organs which is elevated. Also as the organs are inseparable from the whole, it is the man himself who is elevated when any organ is said to be so. The man, as we have seen (n. 23), consists of three degrees or planes, and in each degree all the organs are extant according to the quality of that degree. Moreover, in the natural degree in a defined image a representation of the three degrees is effected, whereby the man can representatively, before he dies and enters the reality, live as a celestial man, as a spiritual man, or as a spiritual - natural man. This is effected by the might of the superior correspondences inducing their images and likenesses on the lower degree, and animating it; though it still remains for the man in nature as the natural degree. Thus there is no unmooring of faculties in their elevation, but, by means of the uplifting, stability and fixation to their USES. Perpetual and varied elevation is the use of the intellect or understanding, and for this purpose it lives above the heart until the heart is at one with it; and when this comes to pass, they, which are the man, are elevated together. When they are elevated there is a new heart and a right understanding,—each faculty in a degree above its original nature. But the higher degree so lives in the lower that no displacement can be thought of, but fulfilment of love : the lower simply becomes what by creation it was meant to be. It becomes a nearer part of the soul's world, the spiritual world, and yields right of way through it to heavenly thoughts, affections and actions. The integrity of the man demands this; for otherwise he becomes inverted, and his head and his feet change places. The actions, interactions and reactions in the spiritual mind according to the ways of its organs, have been luminously expounded by Professor Tafel in his great Work, *Swedenborg and the Doctrines of the New Church*,[1] in which the doctrine of elevation or enlightenment is especially made prominent: and to that work I earnestly recommend the attention of the reader who is willing to pursue this subject.

Psychical Epochs.

236. Our human life is fluid in the hands of our souls and minds, its real Potters. Embryology, in which the little pre-creation is seen to pass through a succession of changes, suggests the plasticity of the fœtus as an easy instance of the transmutation of forms. It has been made to support the supposititious stages from primordial protoplasm into man-

[1] Pp. 571, Speirs, London 1889.

hood; and as the embryo passes, it is fancied, through a whole animal kingdom of development, it is possible to conjecture that the stages can be separated, and are separated in nature, and that ascidian, fish, tadpole, batrachian, and the rest, are distinct progressions one from another, reaching their climax in monkeys and in men. Are we, then, still in a fœtal state in the womb of things, and is every attainment going on and growing on into something else than itself? Embryology might seem thus to suggest a vivid image of evolution. Regarded as a seed without a sower or a soil, *certes* an early fœtus is a loose horse for your formative fancy.

237. If any embryo were outside, and not in a mother; also if it were not an heir but a founder in itself; if it were a self-made embryo in a self-made laboratory, evolution might be illustrated by it. For it might stop

at any particular stage of animal, or travel onwards to any other. But man and wife and their commanded union are the embryo of the embryo, two large factors, and sacred owners of the sacred embryo. It is straitly going to be what they are, and save by death halts at no stage short of one of them. It cannot pause in midway any more than a ray of light can stop unless arrested. So it descends, devolves, from these parents. The pressure of their natures gives it all their organs, mental and bodily. It has a soul very like theirs, and a form and shape. And everybody knows that from the instant of conception it travails and travels into man or woman. This excludes embryology as a curriculum that can be broken or interrupted, and furnish parts for a fish stage, for a frog stage, for an ape stage, or for any other than the goal called a baby, which birth is. Abortion is indeed possible, but it is the end of devolution, and not the beginning of evolution.

238. The human form, always with a divine or human precedent, is the most inviolable of all forms and things. It lives from its own essence or soul, and while that essence remains intact, everything else may be lopped from it, and the man survive. Witness accidents and surgery. "Widdrington in doleful dumps,—when his legs were off he fought upon his stumps." The willsoul can have a project of courage-limbs which are beyond the fates.

239. In short, the human form in the generations is immortal and indestructible even in this world, and in a real sense invariable. It is a type inherited by nature from a divine archetype which conserves it.

240. Yet Swedenborg has been instructed to show that the psychical man has first altered himself, and then has been altered

miraculously, with the great religious epochs. First, to use Swedenborg's own words, there was "homo protoplastus,"—man protoplast, —a condition in which man was as it were an animal, with an apparatus and guidance of human instincts fitted to and widening with all his first wants on earth. His privilege and even then humanity was a Freewill, but with no fixation in it, because he had no habits, and no inheritance of ancestral depravity. He was indeed in the *degré brût*, as near as possible to an animal without being one, for his freedom precluded it. His lowness was the necessity of a responsible commencement ; of a personal opportunity to rise, or not to rise, without which he would not have been a free man. It was a state of human nature or "simplism" in respect to that for which it was the preparation, and which was at hand.

241. In the fulness of days this first natural man, as he freely chose, surceased, and Man

was born. He was born mentally, and by the only possible way,—yes, the only possible; by divine angelic instruction communicated sensibly to his innocent inchoate mind; which instruction he received and appropriated. He thus became a conscious inhabitant of both the natural and the spiritual worlds. This raised him by inclinations and degrees to the celestial state in which a perception of the nature of his teachers and the sanctity of their ministries reigned. The change in the human mind and brain from the lowest natural degree to the celestial was immense. The natural man was born from below upwards apparently, yet God was his Father there; the celestial man was born manifestly to himself from above downwards. Manifestly, for Jehovah by angels visible and invisible put the celestial life into him; the birth, that is to say, from heaven. The new infant impregnating the adult natural man, the child who was this man's father, was thus celestial, and made him celestial. For as this was the provided end, and the adult natural was the divine oval for its field, this womb of flesh had the babe of the regeneration slumbering within it, waiting to be born and called up into life by God and His Angels. Seeing and hearing, hearkening, obedience, and perception, were the protoplasm of the new creature who became of the nature of celestial love.

242. Inward plasticity of mental form, to all substance substantial, and with Eden circumstance environing, is predicated of the human life so far. Did it answer to our embryo state? Surely no. The present embryo is the case of a new soul born by inheritance into a basic flesh of ancestral evils. Moreover, the first preadamite men were not born into the womb, but ovally, into a providential plane of the world. God their

Father had created ovarian climates down which they came, and uterine climates into which they were introduced. They were born of Him as a Father, and of heaven immanent and bowed down into Nature as a general Mother. Thus what is our antenatal state now, with the first natural man becoming instinct with the celestial man, was open world and open heaven, and not confinement in a previous womb. So the first men and women breathed, which embryos cannot do. But the breath, in proportion as the men were lifted by regeneration or second birth, became internal. It was not inspired and expired by themselves without conscious co-operation with the supreme Powers; for they breathed with their angels from altitudes above themselves. It amounted to this,— the mind and the brain drew ends, purposes, principles, perceptions, thoughts, from heaven first, and the heart, the lungs, and the brains translated these celestial things into breaths. So all respiration came from within, and was what Swedenborg denominates Internal Respiration. It came also from part after part of the soul according to its affections, and proceeded into the corresponding tracts of the body, and governed the senses and perceptions. Let your purest love give you some idea of the blessedness of a palpable unition with those above you in heaven ; and your happiest rarest moments of insight, some imagination of the lucidity of a human nature in righteous accord and vision with the Father of Nature.

243. Of all this period of the Golden Age we know only from Scripture and its internal sense; from Swedenborg's commanded revelations ; and by correspondences now again revealed concerning the states of the body before and after birth. Science without learning at these sources tells us nothing. Much, however, can be known, doubtless up

to every man's capacity of knowledge. The traditions of nations represent in evening redness the great morning behind the present Mankind.

244. So soon as man fell, his mental and bodily organization was changed, and the first inspiration and expiration from the soul to the body, breath by breath was given up. Instead of breathing in freedom, that is, by choice, from his will and understanding coupled to heaven, he breathed still in freedom from his new but evil and subverted proprium. He took his own way more and more. His spirit or respiration became external because from himself alone : for it came from without inwards and from below upwards, from his sensual basis. A reversal and retorsion of his mental and bodily organization gradually took place. Primeval man as a living soul was dying. The channels of his inflowing life were one after another

closed and blocked. The body was solid where it was once fluid, and fluid where it was once solid.

245. The end came organically, as it always does in spiritual things, at last. The first breath which Jehovah God had breathed into the perceptive nostrils, and which made man into a living soul,—the spirit of the divine love with its heavenly wisdom,— ceased to be received in any corresponding human love. The heart of self-love alone breathed and thought through the character, and the celestial man was a dead life.

246. The Deluge is the name and representation of this Catastrophe of human nature. Man now hated the first breath with its origin· from Jehovah God: his love of his own acquired nature with its "direful persuasions" stifled it. He had as yet *no other breath*, and his persuasions as great waters suffocated him. His perceptive intellect was drowned

in his lusts, even as man's common-sense understanding may be drowned now. This epochal event is told in correspondences in Genesis, and in full intellectual series by Swedenborg in his *Arcana Cœlestia*, where it stands in crystal cliffs of revelation to be seen afar from age to age.

247. Wherefore, that His child, man, might still exist in both worlds, God altered the human spirit bodily and mental in those men who were capable of sustaining the change; namely, in the remainder called Noah. He opened the lungs to external respiration; so that the man himself, by divine permission, became the motor of thought, intelligence, and corresponding breath; and internal respiration, man's breath from heaven, was closed. In this act, the movements or thoughts of the lungs were so separated from the heart's pulses, that they should be capable of spiritual elevation above

the heart, and stand in front of it, see for it, and guide it. They were, however, united to the heart, as the understanding, whether obeyed or disobeyed, is inseparable from the will. But of this, more presently.

248. This new devolution by the Lord's mercy perpetuated the human race. It met evolution coming up from below: man's merciless self-love making all his faculties into its own image and likeness.[1] It recognised his state, and came to him from the outside air where his thoughts and senses lay, in an external Revelation, in a Word of God not written on the now destroyed heart ; in a record Scripture, for a quite new beginning of an external independent understanding. The first angelic instructors of the Adamic Church were forgotten : their

[1] The monstrous genderings of the ruined Adamic Church are represented by Chaldean Berosus ; see my little Work, OANNES, pp. 16–28 ; where also the celestial state and its internal respiration are treated of.

ministries were now exchanged for a written Word, having continuity of correspondences with the former unwritten Word, and prophetic all through of the Word made Flesh, of the divine humanity and His personal supernal Reason.

249. Did this great revolution in human faculties in any way subvert that first image and likeness of God into which man was created? Was the ground-plan of our nature altered by it? It is a deep inquiry; and we can touch it only on the surface. The change took place in the spiritual organism, in the closing up of the first will now corrupt, and in the abolition of its perceptive but now merely persuasive intellect which saw only what the lusts of the will chose. This with its breathings forth was sealed away from the subsequent humanity, and mankind was emancipated from its fatal influx. But how was the human form, the

vessel of the divine, saved? There are two fountains of its creation : the first by Jehovah Himself; the second also by Jehovah, but through the heavens and their human form. The first humanity is not man's but of God in man. The second is the Man's, and he has his Freedom and Reason, his powers, within it. This he corrupted and destroyed. But the prime humanity was beyond his reach, incorruptible and indestructible. His soul's whole personality lay there. So were there two organisms : a domain of God, and a domain of Man. The latter was saved by the closure of the corrupt will; by a new mind given through understanding, intellect, truths, and through a new affection for them, which affection was raised into a new will. This took place only through the separation and emancipation of the intellect from the domination of the will in what we have called the domain of the man, so that he is no longer a being of inspiration, but of thought and consideration. On the other hand, his will and his intellect are of necessity, as Swedenborg says, "united in all operation and in all sensation." How this is may be exemplified to common understanding both as to the separation of the faculties, and as to their union. A man proposes an action to himself as is his wont continually. In this condition, his intellect perceives the project, whether it be feasible or not, whether it be wise or unwise, good or evil. To do or not to do, that is the question. The thoughts of his mind are separated from the thoughts of his will or heart while weighing the matter. He now decides what to carry out. In the doing and sensing his will and intellect must be united; he must will to do his volition, and see how it is being done; so that his will and intellect are closely together in the resulting operation. Thus there is no severance in his human mind or form at last: in their limits and extreme boundary his breath and his pulse unite in one common eye and muscle and nerve of action.

250. Observe also that as the human form is inviolate by the mercy of the Lord, the lost Adamic perceptive state called the celestial can be approximated to as regeneration advances, by love gaining the ascendant, and the new will being born into the immediate human form, which descends from no angel or spirit, but from the Lord's Divine Humanity.

251. We gather from Swedenborg that the man's love and his wisdom, his will and his understanding, were separate from the beginning; that by this separation he was raised in regeneration from the natural to the celestial degree. The separation was then a constant showing, as the eye is a constant showing to the footsteps, but not an elevation

above the will-faculty, whose way was then from perfection to perfection. For in the climax of that degree his wisdom was the level eye and partner of his love. After the fall, or in the fall, both faculties fell, and the insanity which took the place of the wisdom, completed the selfhood which extinguished the celestial love. Why could not the understanding be separated in that epoch, and elevated above the will, to guide and lead it back to good? As a faculty, it was by creation a separate faculty. The primeval state, the reign of love,—good love and bad love,—is the explanation of it. When either love had become supreme, the intellect was its vassal. There was still separateness of the two faculties, the intellect and the will, but the former worked for the latter and ministered to it. As an illustration now, think of all strong love and the indomitable pretexts of it: how can it not reason? But the

Adamic Man, in his perfection and in his decline, was nothing else than celestial love and its perceptive wisdom, and then self-love and its persuasive insanity, and there was no way to his state, excepting the pressure of heaven which confirmed it so long as it was good, and the rush of hell which confirmed it when it grew evil. This was mere regeneration on the one hand, and mere damnation on the other.

252. Granting this, and that Jehovah God was intimately present, who would not say that His intervention, necessarily miraculous, was at hand? The slave could not be raised into the freeman, for he was a free or most willing slave. The first intellect was ruined and lost. But the "immediate" human form was still there; the heart and the lungs spiritual. The miracle lay, not in the separation of the intellect from the will, for that was already a fact of Creation, but in the

new capacity of elevating the mind, the intellect, above and away from the will, to instruct, correct, reprove, chasten, and reform the will. This capacity, impossible in the celestial man and his descendants, is the essence of the spiritual man as distinguished from the celestial. It is now our possession, and conscience, the new will which arises from it, acting on childhood and its survivals and remains, is the means of our regeneration.

253. Lastly, observe that the human form in all our psychical changes was itself never a ruin. It stands in God: in heaven as the mansion of all goodness and truth and blessedness: in hell as the prison-house forbidding any deeper degradation than the existing character warrants. It is the seal of the Divinity in both estates: the image and likeness of God still extant in the soul.

254. In the celestial man the intellect was raised by the will to its own level, and

became celestial truth or wisdom. In the spiritual man the intellect is elevated from his sensual basis above the will to summon it by self-denials to a like elevation. May we think of how this confessed miracle was accomplished? Surely only by some example of a correspondent miracle. Swedenborg tells us how the dead are raised from the natural into the spiritual world, and that he himself was put through the process for the practical reason of being able to declare the fact to mankind; no other experimental proof being possible for this world. Such resurrection is not a spontaneous act of the incorporate spirit which is bound to the body and its life and automatically cleaves to it. On the contrary, it is a divine visitation and interference. The Lord is the resurrection and the life even in this sense. And our prepared Swedenborg was raised representatively to show it. He was reduced to the condition of the dying, and made sensible of the Divine agency as "a living and strong attraction," which drew the spirit,—itself the real man,—out of the body, detaching it from the intricacies of the corporeal organization, and overcoming the fondness of the nature of the spirit for the nature of the body. May not the dead body of the celestial church be thought of correspondentially from this now public private example? Did not a "living and strong attraction" draw the understanding up and away from the corpse of the will, and summon it to breathe above the suffocating deluge waters in the truth of a new intellectual air; in a New Dispensation? The intellect was indeed depraved like the will, but only in conjunction with the will and while on its level. In itself it is a neutral faculty incapable of depravity, and all blame passes it by, and hits only the will. Separated from the will, elevated above it by

Jehovah, drawn irresistibly and recreatively to the external divine truth which was His new manifestation, and made capable by the elevation of receiving this saving truth, and of becoming amenable to it, a new departure, a New Church, was inaugurated through its means for the regeneration of mankind. This spiritual Rainbow, or new truth, this new organon for man's salvation, took form and body in a new Written Word.

255. There is a passage in Swedenborg's *Economy of the Animal Kingdom* which shows his perception of the celestial state before his mind was opened to the spiritual sense of the Word. He regards and speaks of this state as the *soul* possessing all knowledge, being for the body and the man, Science: every gift of perceiving and using the world constructively. It is unlike the

style of his theological writings, but the reader will do well not to miss it. May I remark here that the above work and the *Animal Kingdom* are mines of Biological and Psychological wisdom?

256. "It appears to be enjoined by the most grave and necessary reasons, that as soon as the soul, which is science, begins to lead a bodily life, it shall cover itself with veils to induce ignorance, and shall only at a late period, or at an advanced stage of life, uncover itself a little of their darkness. For God is a necessary being, and whatever is in God, and whatever law God acts by, is God. If we were born at once in full possession of the perfection and science of the soul, it may fairly and reasonably be doubted, whether the human race could be propagated by natural generation; and whether it would not be most distinctly conscious of its own formation, and by a foregone will overrule

all the details of its growth in the womb, and from the first breath of life continually aim at a more perfect state. But granting that under such circumstances natural birth would be possible, still there is good ground to doubt whether decline and death would be so. In the former case, the earth would not be peopled; in the latter, a thousand earths would not suffice for human prolification. Moreover, in a general state of integrity, there would be a perpetual communication of thought; and therefore little or no speech; and speech indeed could never enunciate what the soul represented to, and beheld within, itself. The soul would look down continually as from a heaven upon its own earth; nor ever cease to raise itself above itself; and then it would require a fresh miracle every moment of its life in the body, to prevent it from exalting itself above God. The least delinquency must be absolutely indelible; and this would give rise to a general perversity and lamentable state, in which there would be no room for grace; because the evil would spring from the very soul as the centre, and not from a mind intermediate between the soul and the body. Furthermore, there would be one general equality between all bodies and souls; consequently no society, either in this or in the future life; for all distinction, and all relation resulting from distinction or difference, perishes in equality. Joy, happiness, good, would not be predicable, because not representable relatively to their opposites. And there are innumerable other consequences besides, to show that it has pleased the Deity that the perfection of the whole should result from the variety of the parts; which variety therefore must be regarded as a necessary means to the ultimate end of creation. Wherefore it is enjoined that the way between the soul and the body should be closed, but to be opened successively as we become adult." (*E. A. K.* n. 299.)

257. As there are survivals of early geological creatures in the existing flora and fauna of the external world, so there are survivals representing to us a likeness of the Adamic or celestial man at the present day, on both the good and the bad side. There are men of the heart and men of the head. Also for the most of us there are moments of the heart and times of the head. And when the quick heart is right, its thoughts are precious beyond other thoughts, and its intellect is as it were again primeval and first-rate. It solves difficulties which are knots to cold reason, and goes straight to other hearts. Its common sense is brilliant. It is the ideal end and means of the deepest sym-

pathies, and the parent of great moments and intuitions, and probably is even now the unknown fountain of the best genius, and the well-head of every real Muse. It justifies a faith in the golden time, now the poem of the world. To this heart our Father appeals as if it were in us still: a divine courtesy, which commands us to love the Lord with all our hearts, and our neighbour as ourselves. But we say this to note that the men in whom this love or heart most reigns are least alterable by outward considerations. They illustrate the state of the Adamic or celestial Church in its inseparable faculties, in less wanting or accepting bare truth than other men, because they have truth alive so near to their hearts of flesh : for truth for them is not truth but good, and they perceive that what is good is also true.

258. On the evil side also there are men of the heart, the most irreclaimable of cha-

racters, in whom the conscience has no place, and to whom instruction comes in vain. The will rules, and the understanding reinforces it, and is not elevated above it. They have no sight, but rush blindly into evil, and for the most part into crime. On both sides, the good and the bad, these are merely representative remainders, and examples that external instructions by truths and their warnings are not so germane to these exceptions as to other men.

The Word in its Epochs.

259. The Word of God, as we have seen, follows the changes in mankind. It was written on the hearts, on the love-centres of the Adamic Men. It constructed the men by regeneration, and taught them the wonders of their new being. After the fall the first heart was lost and closed up, and

another Word was written for the head; for the spirit and lungs of the saved Noahtic men. Inspiration as a general human predicate ceased in them, and doctrine, teaching, schooling was given in its place. The perceptions of the Most Ancient Church were doctrinal records collected and stored for the Ancient Church. When mankind also destroyed the spiritual conscience given in this latter dispensation, ritual revelation came in the Jewish Word to the Jewish mind. A formal ecclesiasticism was the divine remainder, and its observance was religion. When at length obedience itself died out, the Lord Himself came into His creation through a Jewish Virgin. The Word was made flesh. His New Testament was inspired and promulgated. The written Word was complete. Brooding over preadamite man,—that earth without form, and void, and darkness upon the face of its deep,

—and coming through Adam in Paradise, it traverses unwritten and written history, and ends in the Gospels, and in the crowning divine Vision of the Apocalypse with the descent from heaven of the Holy City the New Jerusalem : the prophecy of a new heaven and a new earth in which dwelleth righteousness.

260. This is not progressive Revelation, for it follows man in his downward course through all his stages, and the Incarnation, the " pause of the ages," and the new starting-point, is not of man the crucifier, but of the Lord, the crucified. It is Providential Revelation, the Lord caring for each several necessity of His creature. " He lays His hand upon us behind and before ; such knowledge is too wonderful for us." "Surely goodness and mercy shall follow us all the days of our lives, and we will dwell in the house of the Lord for ever."

261. This adaptation of the written Revealed Word to so many distinct kinds, states and wants of humanity at first and of human nature afterwards, is a subject of comment both to Christians and Critics. It is objected to the Bible that it is a series of small books written in different ages by various authors, and put together as one Book. So it is regarded as a heterogeneous Miscellany. For it consists of Myths, of real histories and fragments of such, of biographical narratives, of rhapsodies and prophecies, of poems, of moral precepts and discourses, and of Visions. For this reason it is either not Divine, not the Word, or the human element so intermixes with it that it is as an imperfect earthen vessel though holding divine contents and instructions. The contents are divine, the form and shape fallible-human ; and unaided except by our own reasons and likings, we have then

before us the stupendous task of clearing the divine element from its obfuscations and disguises. Moreover, God appears in it in different parts of the authorship as a different Being; in the Old Testament and the New with totally different attributes. It is a long subject ; but premising that the uniform spiritual sense from Genesis to the Apocalypse now revealed through Swedenborg demonstrates that the Word is consistently divine throughout, it is at present the object here to answer only the objection of the issue of the Bible by a series of men in different ages as impugning its character as a transcendent divine revelation from the Lord Who is the Word.

262. Now, everything good and evil in the world has come through men. Man, *le monde*, the World, can have no other interlocutors or missionaries. Jehovah, the Ancient of Days, was as an Old Man to the

old Jewish mind. When Christ came He came as a Man. In no other form is appeal to man possible. Angels of Jehovah are always HE and not IT. They are men, and have all of them been men in the natural world. Sinai burning with Jehovah upon it had a missionary Moses to bring down the Ten Tables, and he broke them, and was commanded to write them again himself, so that they were no longer directly from Jehovah, the Divine Man, but in the outward form finitely humanized as the work of a mortal.

263. Much indeed comes not obviously as man to man. All nature comes to him on its own separate ground. But it never stirs the human in him unless his own nature invests it with his own attributes. This happens in Mythology, which peoples the world with quasi-human deities. These, however, though like revelations, and in some continuity of survival from them, have

no power to lift man to God, or power only so far as they approximate to what is nobly human. So also natural sciences are nothing but man's mind with nature lifted into it.

264. But now in what other way than that of a successive issue of Words, could divine revelations to man, applying themselves to his spiritual changes, be given? Would it meet the case of perpetual varying divine instruction called for, if the Revelations were all given at first in one book by some one chosen person, as the Koran by Mahomet; —of course at the beginning of the Days? Would infant humanity, if it had to master it, retain, or survive, such a Codex? The Book must be both a Revelation and a Legislation for all the ages. And its contents would for the most part be premature, and would be with each varying epoch inappropriate, and to the next epoch outworn

and effete; thus more subject to sceptical criticism than the Bible is; and though all of one piece like a theologic Justinian Code, it would surely be objected that it showed only the mode and mannerism of one narrow inflexible author who from this circumstance was not God after all, but a pedantic pretender. The critics would then cry aloud for what we now have,—for other utterances to balance the idiosyncrasy of such a Bible; for Revelations suited to times. But how if Jehovah were Himself to promulgate it, and place it down in the ages? Would it not then be as the primordial cell of the scientists, and evolve religion out of itself for every opportunity? Yes. But if the critics were "fanatics for evidence," and as strong natural men denied Jehovah and His authorship, soon become a tradition, He must, as Renan postulates, repeat His Word and His Work over again often at stated times at the call of learned professors establishing divine truth by vote. In this case we fall again into the way of successive issue of revelations, though not as needed by man, but as demanded by scientist minds. The human exigency, and divine supply, is indeed met on Sinai-mount, and repeated all through the Word; and it follows the lines of that great common sense, that "man's extremity is God's opportunity."

265. Therefore the matter is reasonable as it stands. Admit that God supervises the world. Admit that history and personal experience show it as a changing world, and that a Divine Providence oversees and measures the changes. Admit that Canaan, Judæa, Babylon, Assyria and Egypt are different in genius and mind from the world after Christ ; and that the modern world since the middle ages is diverse from the ancient times, and especially that the eighteenth and

the nineteenth centuries with their eager nations are different again: and humbly and reverently thinking of our Father Who is in the heavens as our divine schoolmaster, we may begin to imagine that a succession of school-books will be likely to be given out through heaven for our instruction. The Books for ancient Egypt versed in talking by hieroglyphical representatives, would not suit the genius of modern England, or Germany. But we need hardly pursue this tale of difference.

266. So is not a succession of revelations suited to our race in its spiritual changes what we should expect? Every man and woman in common education goes through successive phases of teaching and teachers; and why not the educable race in things divine?

267. The real objection is to authoritative divine instruction to be accepted with no criticism and no after-thought. Such instruc-

tion must be miraculous, and miracle offends. Yet miracle must come and will be stated. Miracle is not to convince or dominate unbelievers, but to found religion with the simple who so love, and lead, good lives, that new ages can be born from them when old ages die. Christ did no miracles in one place "because of the unbelief" there. He did miracles elsewhere because of the belief. The evidence of all miracles is immediate and personal, and the transmissibility of the evidence involves the same humble capacity as the first witness of the miracle. Things hidden from the wise and prudent are revealed unto babes.

268. We come to the safe conclusion, that the issue of fresh Revelations from Almighty God through chosen men,—through oracular persons writing of spiritual creation in correspondences as in Genesis, chap. i.-xiii., through inspired historical scribes, through

Lawgivers, Prophets, Psalmists, Seers, Evangelists, through men of divine Apocalyptic vision,—are a series called for from the first to the last by the human occasion, and provided by the divine mercy. And when we know by the Revelation of the internal sense through Swedenborg's Commission and complete Work that the writings of all these functionaries are one Book by One Author; and that that Book is the eternal Word, and that Author the Lord, we find that the succession of books is swallowed up in the unity, and that God's ways are justified to man.

269. Thus Revelation seen internally is the seamless garment which the soldiers do not part. Consisting as it does of spiritual and natural Correspondences cast into a divine letter, it belongs to all ages as creation itself belongs. Light and Heat signified wisdom and love for the Adamic Church as they signify it for the New Church. Heart and

breath from the beginning signified and
signify will and understanding. The human
form and all that it is for ever corresponds to
the provinces of Man in heaven. The Earth
and all things in it, gardens, trees and living
creatures, piece for piece, have spiritual souls
of meaning which make nature essentially
into a theatre representative of the Lord's
Kingdom. So the Word is written in a
divine language. This as a firmament spans
the ages. It is exactly hieroglyphic; the
Word is its speech: the world,—sun, moon
and stars,—is its natural realization: the
Church is its depository. It belongs to the
primeval childhood of the race. It was
spoken by the Lord when the Word in Him
became flesh. It belongs to childhood for
ever, and fathers and mothers can build up
the True Christian Religion from the earliest
intelligence of their sons and daughters
through delight and wonder and instant

belief by elements that are as mother's milk for nourishment, and that change into faculties that are the forms of adult spiritual life. For the correspondences of the Word are born into the mind like children, and grow up into men and angels for ever and for ever. Thus the Word unites the ages of the individual and of the collective man into one great age which is the image of Him who is the Alpha and the Omega, who is and who was and who is to come. This is its omnipotence on earth. But the same correspondences exist in heaven, and effect a similar unity. The hosts above live in the perception of them, and think them there ; yea love and think them into visible existence. Heaven and Earth therefore, separate in appearance, separate as discrete planes, meet in these precincts of the supreme Mind, and are united in the Word. The grand man, the little child, and our historic and present race, are equally in its tutelage ;

and this, through the divine language and life of correspondences. To the simple they are unconsciously, may we say automatically operative while they read the Word : they impart a holy feeling which keeps evil spirits away, which blesses the day's work, and sanctifies the life. To those who are otherwise competent and disposed, they must be sedulously learnt, and the doctrines which gather them into spiritual good and truth, and thus form them into faculties of a new will and understanding, must be steadfastly affirmed. Gentle Reader, open your heart's mind to these Correspondences, and in its warmth contemplate them, and you will be astonished at the contents of the Holy Word.

POSTSCRIPT.

SINCE this book was written, I have seen a work by the late John Spurgin, M.D., F.C.P.S., namely, *Six Lectures on Materia Medica and its Relations to the Animal Economy*, delivered before the Royal College of Physicians in 1852 (London : John Churchill, 1853). The sixth Lecture contains a luminous statement of Swedenborg's theory of the Coronary Vessels of the heart, with extracts verbatim from *The Economy of the Animal Kingdom:* also with original confirmations by Dr. Spurgin himself. He brings forward a case from *The Dublin Hospital Reports for* 1827–1828, in which the so-called coronary arteries were not only ossified, but their cavities completely closed for an inch in length from their [supposed] origins, so that for the last inch they had no perceptible canal. "In this case," Dr. Spurgin says, "it is manifest that the heart received no blood from the aorta." He says further: "According to the generally received opinion, as well as according to the appearance, the parietes of the heart are supplied with blood from the coronary arteries which first arise from the aorta. But if the fact is so, then the heart, for its supply of blood, is dependent upon the arteries, and the arteries again are dependent upon the heart. Which is the primum mobile, the coronary arteries, or the heart? Muscular fibre cannot act without blood, and the artery which supplies the blood cannot possibly receive that blood without the action of the heart." Dr. Spurgin reiterated his strong conviction concerning the coronary vessels in his HARVEIAN ORATION.

He was the first English physiologist to accept Swedenborg's view; which was propounded in the work mentioned above, in 1740. It gives me pleasure to refer to this subject, because to the Doctor I owe my earliest interest in the Anatomical, Physical and Philosophical Works of Swedenborg, the translation of which has been the most important occupation of my literary life.

www.ingramcontent.com/pod-product-compliance
Lightning Source LLC
Chambersburg PA
CBHW021213240426

43667CB00038B/632